FUNDA   TALS
OF   TY
FOR   RE

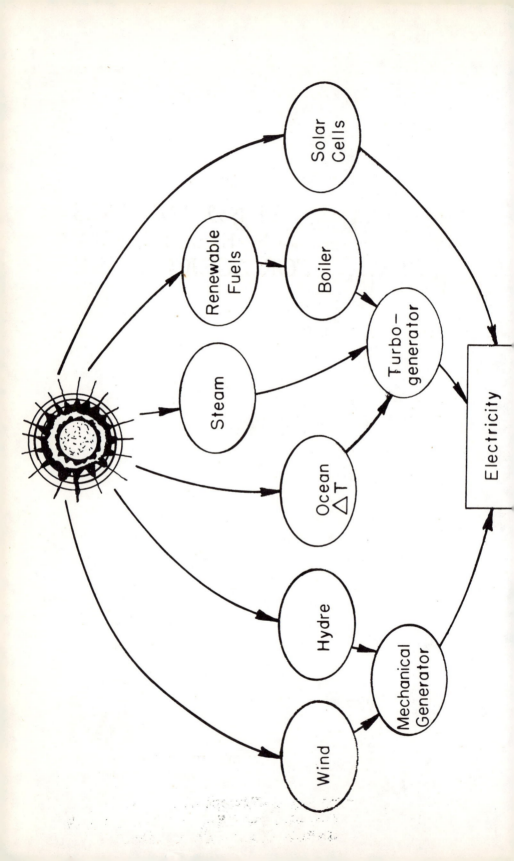

FUNDAMENTALS
OF ELECTRICITY
FOR AGRICULTURE

Robert J. Gustafson

Associate Professor
University of Minnesota

AVI PUBLISHING COMPANY, INC.
Westport, Connecticut

Library of Congress Cataloging in Publication Data

Gustafson, Robert J
  Fundamentals of electricity for agriculture.

  Includes index.
  1. Electricity in agriculture. I. Title.
TK4018.G87 621.3   79-13443
ISBN 0-87055-327-5

Printed in the United States of America

Preface

One of the most versatile and widely used forms of energy is electrical energy. Our homes and our agricultural operations have become dependent on electrical energy. In 1927 less than 3% of the farms in the U.S. had electricity. Fifty years later 99% of all farms had electricity with an average electrical use of over 1300 kilowatt hours per month. Because agriculture has become reliant on electrical energy and our use of electrical energy in agriculture continues to grow, it is imperative that we understand the basic principles of electricity and develop the ability to apply them to practical situations.

This text is designed to assist the student both in attaining a basic understanding of the nature of electricity and developing skills in solving the problems associated with applying electricity in agriculture. The book does not assume any previous knowledge of electricity. It is arranged such that the first portion lays a foundation in basics of electricity. The second portion of the book deals primarily with applications and devices commonly found in agricultural uses of electricity.

This text adheres to the philosophy that mastery of the understanding of how and why things work, in addition to skills in manipulating the simple formulas and procedures in problem solving, is necessary to successfully apply electricity. Through such understanding the student will be prepared to cope with the non-routine problems and new developments that are encountered in real life situations.

The broad spectrum of topics included in the text ensures the student at least an introductory understanding of the aspects of electrical and electronic technology and their language that he will encounter in his work. At every step fundamentals are stressed and problems are set in as real a perspective as possible, to strengthen the student's potential to analyze and solve problems not specifically studied in the classroom. Within appropriate sections, the economic implications are integrated with technical factors, in order to make students aware that in real-life applications, costs are a significant influence. Extensive use is made of example problems to demonstrate concepts in this text. Both the International System (SI) and English systems of units are used. It is felt that English-dimensioned tools and products will remain in use in decreasing quantities for a significant period of time. However, most new products that the student will encounter will make use of SI dimensions. Therefore, it seems necessary to obtain a working competence in both systems. To assist the users, Appendix A lists conversion factors for the two systems. Reference to the appropriate sections of the National Electric Code are made throughout the application section.

The Appendices are constructed to include all the materials needed to solve the exercises included in the text. The application and appendix sections may be useful for other courses and in professional practice.

This book is not intended to develop such skills as wiring methods. Practical laboratories associated with the course are highly recommended to reinforce basic principles and bring concepts into reality.

May 1, 1979   Robert J. Gustafson

Contents

Basic Terms and Definitions

1.1 ATOMS AND ELECTRICITY

Early experimenters and inventors established certain basic principles and concepts concerning the behavior of electricity. However
additional knowledge of the physical structure of matter has allowed
modern man to develop many of the electrical devices in use today.
It is therefore necessary to begin our study of electricity with a
review of some of the basics about the structure of material as they
relate to electricity.

We can best describe the electrical nature of matter by reviewing
its structure, starting with elements. An element is a substance which
cannot be decomposed to form simpler substances by any chemical
means. The smallest portion into which an element can be subdivided
without losing its physical and chemical properties is called an atom.
One concept of the atom is based on a model described by the
Danish physicist Niels Bohr in 1913. This model describes the atom
as spherical in shape with a small core or nucleus in the center sur-
rounded by a number of spherical concentric shells. The nucleus
contains positively charged particles, protons, and electrically un-
charged masses, neutrons. The electrons, negatively charged masses,
are distributed in the shells of the atom. The number of electrons
and protons are equal, thereby producing an overall neutrality.
Figure 1.1 shows the helium and lithium atoms.

The diameter of a typical metallic atom such as silver is approximately 3×10^{-7} mm. The diameter of the nucleus is approximately

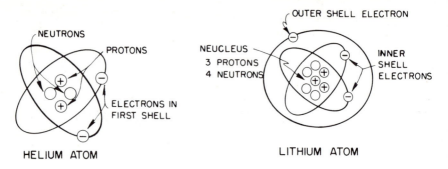

FIG. 1.1.   HELIUM AND LITHIUM ATOMS

10,000 times smaller than the whole atom, a point at the center for all practical purposes.

According to modern theories, electrons in an atom can possess only certain discrete amounts of energy, and therefore must move around the nucleus in predetermined paths or energy levels. The energy levels of electrons in any one shell tend to be nearly the same because the distance from the nucleus is about the same. The energy levels for a shell are referred to as its energy band. Energy levels increase as the shells get further from the nucleus. Between shells are forbidden gaps, energy levels where electrons cannot remain. Energy is required from an outside source to move an electron from one shell through the forbidden gap out to another shell. Conversely, energy is given up by the atom if the electron drops into a lower energy shell.

The electrons in the shell farthest from the nucleus, called the valence electrons, experience a comparatively weak attraction to it. It is the valence electrons which control the electrical and chemical properties of the atom. The valence electrons may wander into outer shells of nearby atoms. Random wandering of outer electrons from one atom to another does not produce any permanent change. If no outside influence disturbs the balance, the overall material remains neutral. If some outside force, such as a battery voltage or application of heat, disturbs this balance, the loosely-bound outer electrons may tend to move in one direction. The result of this nonrandom drift of electrons is called *current* in the line of the drift. Electrical current is found to propagate along the conductor at the speed of light (3×10^8 m/sec), a much higher velocity than the average random drift velocity.

Current is measured in terms of units of electrical charge per unit time that pass a plane perpendicular to the direction of current flow. The greater the amount of charge that flows per unit time, the greater the current.

$$\text{Current} = \frac{\text{Charge}}{\text{Time}} \quad \text{or} \quad I = \frac{Q}{T}$$

where
$$I = \text{current in amperes}$$
$$Q = \text{charge in coulombs}$$
$$T = \text{time in seconds}$$

One coulomb is equivalent to 6.2×10^{18} electrons. Therefore, that number of electrons passing in one second is defined as one ampere.

$$1 \frac{\text{coulomb}}{\text{second}} = 1 \text{ ampere}$$

The term ampere is in honor of the French scientist Ampere (1775–1836) who laid the foundation for electrodynamics.

It should be pointed out before continuing that there exist two conventions regarding the direction to be assigned to the current flow in a network. The majority apply the *conventional* current flow approach, which is defined to be the opposite to that of electron flow. We today know negatively charged particles, electrons, actually move in the material. However, at the time electrical energy was first studied it was thought that positive carriers moved, hence the reverse convention on current and movement of electrons. The conventional system will be used throughout this text.

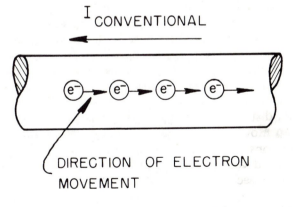

FIG. 1.2. CURRENT FLOW CONVENTION

1.2 RESISTANCE TO CURRENT FLOW

The relative difficulty with which current can be transmitted in a material is defined as the electrical *resistance* of the material. In some materials, such as metals which have large atoms and very orderly structures, the electrons in the outer shells have low resistance to movement; these materials are referred to as *conductors*. At the other end of the spectrum are such materials as rubber, glass, bakelite, and mica which do not readily release their outer electrons to form current. They offer a high resistance to electron flow and are termed *insulators*. In a good insulator, a large amount of energy is required to move a valence electron.

An intermediate group of materials, known as semiconductors, offers an intermediate resistance to electron flow. Semiconductors are very important because they form the materials used in construction of solid state devices such as diodes and transistors. Solid state devices will be considered in detail in a later chapter.

1.3 ELECTROMOTIVE FORCE

Since the useful applications of currents depend on work they can do, it is important to understand how current does work. To do so the fundamental concept of electromotive force is required.

Electromotive force is the property of a device that supplies the energy for movement of electrons in a circuit. The unit of electromotive force is the volt and has the letter symbol E or V. Voltages may be generated by means of mechanical, magnetic, pressure, thermal, radiation or chemical effects. Unlike current which has a flow characteristic, voltage is a difference variable in the sense that two points in an electrical system may have a difference in voltage level. Voltage is often likened to the pressure that causes the flow in a fluid system, whereas voltage "causes" charge to flow in an electrical system.

As an example, a typical car battery has two terminals having a potential difference of 12 V between them. The term potential difference is derived from the concept of potential energy, energy a body possesses by virtue of its position. The difference in potential is established between the terminals at the expense of chemical energy reactions within the battery. The positioning of the charges at the expense of chemical energy will result in a flow of charge (current) through a conductor placed between the terminals.

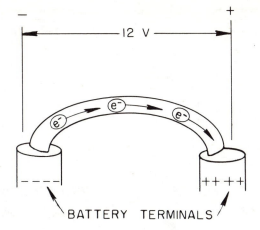

FIG. 1.3. BATTERY FOR PRODUCING POTENTIAL DIFFER-
ENCE

The potential difference between any two points, 1 and 2, in an
electrical system is determined by

$$E_{12} = \frac{W_{12}}{Q_{12}}$$

where E_{12} = potential difference in volts

                W_{12} = energy expended or absorbed in joules

                Q_{12} = charge measured in coulombs

W_{12} is the energy expended (or absorbed) in moving charge Q_{12} from
point 1 to point 2. Therefore, by definition, if one joule of energy
were required to move one coulomb from point 1 to point 2, there
would exist a potential difference of one volt between the points.

The basic difference between current (a flow variable) and voltage
(a difference variable) affects the measurement of each. The voltmeter does not break the circuit, but is placed across the element for
which the potential difference is to be measured. In contrast, for an
ammeter the circuit must be broken to place the meter in line thereby determining the flow in the circuit. (Figure 1.4).

1.4 POWER AND ENERGY

If an electrical charge is carried through a difference of potential
(expressed in volts), work is done. If a certain quantity of electrical

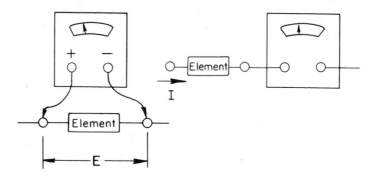

PROPER VOLTMETER
CONNECTION

PROPER AMMETER
CONNECTION

FIG. 1.4. VOLTMETER AND AMMETER CONNECTIONS.

charge is moved from one point to a second point at a different potential, work is done. If accomplished in time, t, the rate of work or *power* can be determined. Current has been defined as the rate of flow of charge; therefore the product of the potential difference and the current is power. The following dimensional analysis helps to explain this concept.

$$\text{Voltage}\left(\frac{\text{Work}}{\text{Charge}}\right) \times \text{Current}\left(\frac{\text{Charge}}{\text{Time}}\right) = \text{Power}\left(\frac{\text{Work}}{\text{Time}}\right)$$

The common unit of power is the watt. Power in watts is given when the potential difference is in volts and the current in amperes.

Power (watts) = Potential (volts) $\times$ Current (amperes) = EI

Example 1.1.

Calculate the power from a 10 amp current caused by a 120 V potential.

Power = E $\times$ I = 120 V $\times$ 10 amp = 1200 W

We have defined power in terms of voltage and current. Recalling that power is the time rate of energy use or work, we can express electrical power in terms of the energy unit of the joule.

$$\text{Power (watts)} = \frac{\text{Energy (joules)}}{\text{Time (seconds)}}$$

Example 1.2.

What is the energy required to operate a 300 watt water heater for twenty min?

Energy = Power × Time = 3000 W × 1200 sec = 3,600,000 joules

Power is measured by an instrument called a wattmeter. The wattmeter gives a reading of the product of the current and voltage in a circuit. It has a minimum of two terminals for voltage sensing and two terminals for current measurement. The voltage terminals are attached across the load (as a voltmeter would be) while the current terminals are connected in line with the load (as an ammeter).

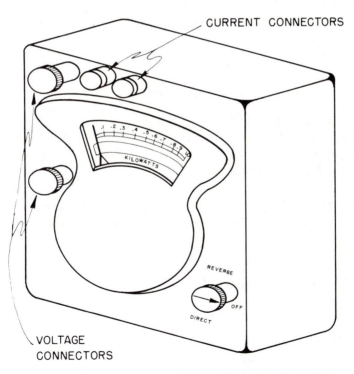

FIG. 1.5. WATTMETER FOR ELECTRIC POWER MEASUREMENT.

FIG. 1.6. KILOWATT-HOUR METER FOR ELECTRICAL EN-
ERGY MEASUREMENT.

Total energy use over a period of time or power use is measured by a meter called a watt-hour or kilowatt-hour meter. This type of meter is used for billing customers for electrical usage. This device sums the energy consumption over time and gives either a digital or dial display of the total.

The number of kilowatt-hours is determined by the expression

$$kWh = \frac{PT}{1000}$$

where
P = power in watts
T = time in hours

Example 1.3.

Energy Cost Determination.

Determine the total cost of using the following appliances for the amounts of time indicated. Assume the cost of electrical energy is three cents/kWh.
a. 1000 W heater for 3 hr
b. Six 60 W bulbs for 10 hr
c. 2000 W motor load for 2 hr

SOLUTION

$$kWh = \frac{(1000 \text{ W} \times 3 \text{ hr}) + (6 \times 60 \text{ W} \times 10 \text{ hr}) + (2000 \text{ W} \times 2 \text{ hr})}{1000}$$

= 10.6 kWh

Cost = 10.6 kWh $\times$ 3 cents/kWh = 31.8 cents

1.5 RESISTANCE AND OHM'S LAW

The two quantities of voltage and current can be related through the physical parameter, resistance. The voltage supplies the potential force in an electrical system. Flow of charge or current is the desired result. The level at which an electrical system is able to accomplish the above is determined by the resistance of the system. The greater the resistance to flow of charge, the less the resulting current of the system. This is expressed in *Ohm's law* as

$$I = \frac{E}{R}$$

where
I = current in amperes
E = potential in volts
R = resistance in ohms

Example 1.4.

Determine the current flow in an electric heating element having a resistance of 10 ohms if the applied voltage is 120 volts.

$$I = \frac{E}{R} = \frac{120 \text{ V}}{10 \text{ Ohms}} = 12 \text{ amp}$$

Example 1.5.

Determine the internal resistance of a time clock that draws 0.2 amp at 120 volts.

$$R = \frac{E}{I} = \frac{120 \text{ V}}{0.2 \text{ amp}} = 600 \text{ ohms}$$

Combining Ohm' law and the earlier expression for power, we obtain the following important and useful relations

$$P = EI$$

$$P = I^2R$$

$$P = \frac{E^2}{R}$$

where P = power in watts
I = current in amperes
E = potential in volts
R = resistance in ohms

Example 1.6.

If an appliance has a nameplate rating of 1200 watts at 120 volts,
a. what is the resistance of the appliance?
b. what will be the current flow through the appliance?
   a) Using the third relation

$$R = \frac{E^2}{P} = \frac{(120 \text{ V})^2}{1200 \text{ W}} = 12 \text{ ohms}$$

   b) Using the first relation

$$I = \frac{P}{E} = \frac{1200 \text{ W}}{120 \text{ V}} = 10 \text{ amp}$$

The resistance of wire used in electrical conductors is dependent on the nature of the material and the physical dimensions. Resistance is found to be directly proportional to the length of wire and inversely proportional to the cross sectional area. This can be expressed as,

$$R = \rho \, \frac{L}{A}$$

where

R = resistance
L = length
A = cross-sectional area
ρ = resistivity

The resistivity of a material is the resistance of a specimen of unit length and unit cross sectional area. Table 1.1 gives resistivity values for some materials commonly used as conductors.

An examination of wire tables shows that diameter of electric conductors is commonly given in mils. A mil is 0.001 in. The area of conductors is given in circular mils. This is *not* the area $\pi d^2/4$. It is the diameter of the wire in mils (thousandths of an in.) squared. A circular-mil-foot, as used in Table 1.1, is the volume of a wire 1 mil in diameter and whose length is one foot.

Most conductors increase in resistance with increase in temperature.

TABLE 1.1.   RESISTIVITY OF SOME COMMON CONDUCTORS

Material	Ohm/cm^3	Ohm/cir-mil-ft
Annealed copper	67,140	10.37
Hard drawn copper	69,800	10.78
Hard drawn aluminum	110,000	17.00
Iron	375,500	58
Nichrome	4,270,000	660
(Ni-Fe-Cr alloy)		
Silver	64,000 to 72,500	9.89 to 11.2

The temperature coefficient of resistance of a conductor is defined as the change in resistance per ohm of resistance per degree change in temperature from some reference temperature (usually $20°C$). Hence resistance can be expressed,

$$R_t = R_i (1 + \alpha \Delta T)$$

where R_t = resistance at specified temperature

R_i = resistance at reference temperature ($20°C$)

α = temperature coefficient of resistance at $20°C$

ΔT = difference in temperature between specified and reference ($\pm °C$ from $20°C$)

Table 1.2 gives the temperature coefficient for some common conductor materials.

TABLE 1.2.   TEMPERATURE COEFFICIENTS AT $20°C$

Material	Coefficient
Aluminum	$3.9 \times 10^{-3} \dfrac{1}{C}$
Copper	$3.9 \times 10^{-3} \dfrac{1}{C}$
Iron	$5.0 \times 10^{-3} \dfrac{1}{C}$

Example 1.7.

What is the resistance of a conductor at $100°C$ whose temperature coefficient is 0.004, and whose resistance at $20°C$ is 50 ohms.

$$R_t = R_i (1 + \alpha \Delta T)$$

$$= 50 \text{ ohms } (1 + 0.004 \times 80)$$

$$= 50 \times 1.32 \text{ ohms} = 66 \text{ ohms}$$

Tables A.1 and A.2 of Appendix A show the diameter, weight, and resistance for common sizes of copper and aluminum conductors.

1.6 DIRECT AND ALTERNATING CURRENT

Electrical systems are generally classed into two categories by the form of the current. Direct current (DC) is characterized by non-time variant current flow in one direction.

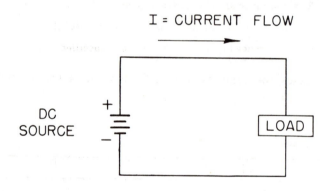

FIG. 1.7.   DIRECT CURRENT ELECTRICAL SYSTEM.

Batteries, thermocouples, solar cells, and rotating DC generators are all examples of sources for direct current systems. On the other hand, alternating current (AC) is characterized by alternating flow in in two directions. Because practically all modern power distribution systems use alternating current, a more detailed description of alternating current characteristics will be given.

An alternating current is one in which the direction of flow changes periodically. The electron movement is first in one direction, then the other. Most commonly the variation is of a sine waveform.

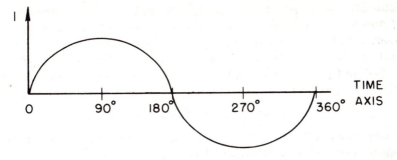

FIG. 1.8.   ALTERNATING CURRENT WAVEFORM

A sine voltage curve is a graph of the equation

$$e = E_m \sin \theta$$

where

e = the instantaneous voltage

E_m = maximum voltage

θ = angle

The instantaneous voltage, e, depends on the sine of the angle. It rises to a maximum as the angle reaches 90°, then falls to zero at 180°. It becomes negative and reaches a negative peak at 270°, then returns to zero again at 360°. In the 360° it has completed one cycle, which is then repeated.

The frequency (f) of an alternating current or voltage is the number of complete cycles occurring in each second. Frequency is generally given in units of hertz (Hz), one hertz equaling one cycle per sec.

1.7 AMPLITUDE OF SINE WAVES

One of the most frequently measured characteristics of the sine wave is its amplitude. In contrast to the DC measurement, the magnitude of alternating current or voltage can be measured in several ways.

One method is to measure the maximum amplitude of either the positive or the negative part of the cycle. The value obtained is called the *peak voltage* or *peak current*. An oscilloscope or a special meter is required to make the measurement.

A more common measurement is to determine the effective or RMS (root mean square) values. The RMS value of a sine wave is the equivalent to the DC magnitude that would provide the same amount of power. For example, many conventional residential wiring systems operate at 120 volts effective or RMS value. This means that a light bulb will glow at the same brightness on the 120 volt RMS or connected to 120 volt DC source. The effective value for a sine wave will be the peak value divided by the $\sqrt{2}$, that is

$$E_{RMS} = E_{Peak}/\sqrt{2} \text{ and } I_{RMS} = I_{PEAK}/\sqrt{2}$$

Unless designated otherwise, reference to AC voltage and current refers to the RMS or effective value.

1.8 PHASE RELATIONS AND POWER IN AC CIRCUITS

When a sine wave voltage is imposed on a load, a sine wave current will result. For the case of a resistive load, the voltage and the current waves are "in-phase" with each other. The term "in-phase" means the current and the voltages go through zero and through the peak values at the same time.

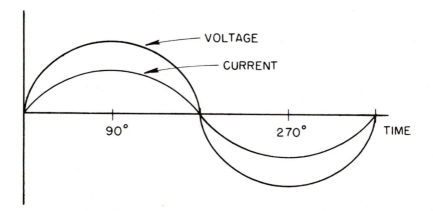

FIG. 1.9.   VOLTAGE AND CURRENT CURVES "IN PHASE"

Instantaneous power in the circuit can be determined by multiplying voltage and current at any instant in time. If instantaneous power is plotted for a complete cycle, we see the power wave form shown in the figure which follows.

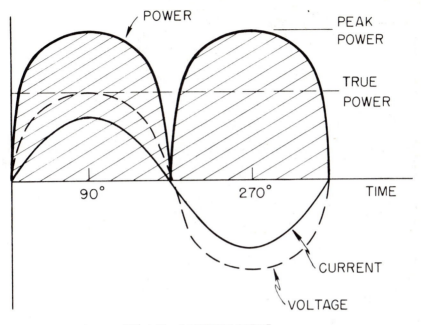

FIG. 1.10.   AC POWER CURVE

True power or effective power as shown on the diagram is the product of the RMS voltage and current. Note that when the voltage and current are in-phase, all the power in the diagram is positive. Positive power is referred to as true, real or in-phase power. The significance of this type of power is that it converts electrical energy into some other form of energy. This will be contrasted with another type of power which does not convert electrical energy to any other form.

When a circuit contains elements with other than pure resistance (capacitance or inductance), a phase shift will occur between the voltage and current waves. That is, the waves will no longer reach their minimum values at the same time. The amount of shift measured in degrees is called the phase-shift angle or the phase-shift. This phase-shift also affects the shape of the power curve. As indicated in the diagram which follows, part of the power curve will fall in the negative area when a phase shift is present. The power below the axis is called wattless or *reactive power*. It does no work. Reactive power can be measured with a VAR meter. VAR means volt-ampere-reactive. One unit of reactive power, one VAR, and one unit of true power, one watt, are numerically equal, but the distinction is that the VAR does no useful work.

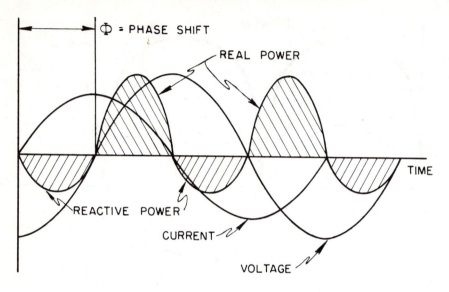

FIG. 1.11. AC POWER CURVE WITH PHASE SHIFT

True power in an AC circuit can be calculated by

$$P = EI \cos \phi$$

where
E = voltage (RMS)
I = current (RMS)
ϕ = phase-shift angle

Note when the phase-shift angle is zero as in a purely resistive circuit, the cosine of the phase shift is one. Therefore for pure resistance we return to the form

$$\text{Power} = \text{Voltage} \times \text{Current} = EI$$

Example 1.8.

Determine the true power output of an AC electric circuit with a voltage of 120 volts, a current of 10 amperes, and a phase shift of 20° between the voltage and the current.

$$\text{True Power} = P = E\,I \cos \phi$$
$$= 120\ V \times 10\ \text{amp} \times \cos 20°$$
$$= 1128\ W$$

In a circuit where voltage and current are out of phase, we may refer to the *apparent power* value as well as the true power. Apparent power is the product of the measured voltage and current in units of volt-amperes. Apparent power is always greater than true power when the voltage and current are out of phase. The ratio of the true power to the apparent power for a circuit is defined as the power factor. The power factor is also equal to the cosine of the phase-shift angle.

$$\text{Power Factor} = \cos \phi = \frac{\text{True Power}}{\text{Apparent Power}} = \frac{\text{watts}}{\text{volts} \times \text{amperes}}$$

Example 1.9.

Determine the power factor and phase-shift angle for a circuit where the true power is measured and found to be 3840 watts, the voltage 240 volts and the current 20 amperes.

$$\text{Power Factor} = \frac{\text{True Power}}{\text{Apparent Power}} = \frac{\text{watts}}{\text{volts} \times \text{amperes}}$$

$$= \frac{3840 \text{ W}}{240 \text{ V} \times 20 \text{ amp}} = 0.8$$

$$\text{Power Factor} = \cos \phi$$

$$\text{therefore} \quad \phi = \text{arc cos } (0.8)$$

$$= 36.87°$$

Phase shift, power factor, inductance and capacitance will be studied in more detail in later sections. As will be explained in greater detail, an electrical system can be arranged to operate with a power factor of 1.0. A system operates at maximum efficiency when the power factor equals one. However this is not always feasible in real systems. Power generators are concerned with power factor because it affects the amount of power they must generate to service the load. Watt-hour meters measure true (in-phase) power usage at the consumer's location. However the power company must generate the apparent power value. Any phase shift created by the load will be reflected in a lower power factor and more loss through reactive power. Therefore some power companies may add penalties for bad or low power factors in their rate structures.

1.9 VECTOR REPRESENTATION

Vector representation of currents and voltages is often used in solving problems involving alternating current and voltage. Therefore a brief development of vectors is in order. A simple vector is a quantity having magnitude and direction. It can be denoted by a straight line drawn to scale to denote the magnitude. Direction is indicated by an arrow at one end of the line in conjunction with the angle the line makes with a horizontal reference line.

Vectors may be rotated like the spokes of a wheel to generate various angles. Positive rotation, positive angles, are generated by the counterclockwise rotation of the vector. Clockwise rotation generates negative angles.

The vertical projection of a rotation vector may be used to generate a sinosoidal voltage form. The length of the vector represents the maximum voltage, E_m. As the vector is rotated, the vertical projection gives the instantaneous voltage.

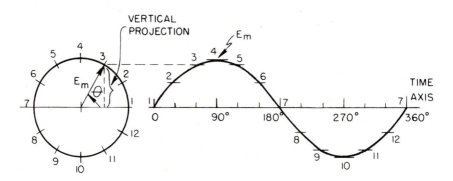

FIG. 1.12. SINOSOIDAL FORM BY VECTOR ROTATION

The trigonometric equation for the sine wave can be expressed in terms of angular velocity of a rotating vector. Angular velocity refers to the angular degrees or radians through which a voltage vector rotates per second. The Greek letter omega (ω) is used to denote angular velocity. The sine wave equation can now be expressed in terms of time as

$$e = E_m \sin \omega t$$

where

e = instantaneous voltage in volts

E_m = maximum or peak voltage in volts

ω = angular velocity in degrees/sec

t = time in seconds

with $\omega t = \phi$ = angle of rotation in degrees

Vectors may be used to combine AC waveforms of the same frequency. The angle between the vectors indicates the phase difference between the waves. The length of the vectors generally represents their RMS value.

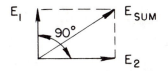

FIG. 1.13. VECTORIALLY COMBINING AC WAVEFORMS

Sine wave E_1 and E_2 are $90°$ out of phase. Wave one reaches its maximum $90°$ before wave two, therefore wave 1 leads wave 2 by $90°$. The two waveforms can be represented by two vectors $90°$ apart. Using the convention of counterclockwise angles positive, E_1 would be $90°$ counterclockwise from E_2. Note the sum of the two waves is a wave of the same frequency as E_1 and E_2 but of a different magnitude.

Example 1.10.

Find the vector sum of two voltage waves of the same frequency and the same voltage (208 volts), if one wave leads the other by $90°$.

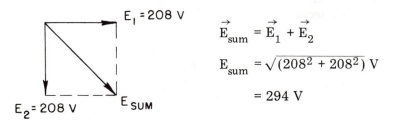

$$\vec{E}_{sum} = \vec{E}_1 + \vec{E}_2$$

$$E_{sum} = \sqrt{(208^2 + 208^2)}\ V$$

$$= 294\ V$$

EXERCISES

1. Using the relationships of P = EI and E = IR, show that

   a) $P \dfrac{E^2}{R}$

   b) $P = I^2 R$

2. Farmer Shock wants to install some 250 watt infrared heat lamps for pig brooding. How many of these lamps can he operate on a 20 ampere–120 volt circuit?
3. What is the operating resistance of one of the above 250 watt lamps?
4. Which 120 volt light bulb has the most resistance, a 100 watt bulb or a 40 watt bulb?
5. What would it cost to use one 1200 watt heater for 10 hr, if electricity cost $0.03/kWh?
6. What current does a 2 kilowatt water heater draw when operating on 240 volts?
7. Find the vector sum of the following voltage vectors.

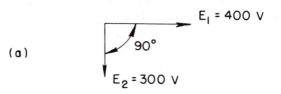

(a)

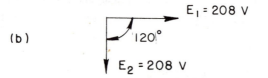

(b)

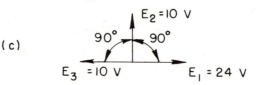

(c)

REFERENCES

BOYLESTAD, R. and L. NASHELSKY. 1976. Electricity, Electronics and Electromagnetics: Principles and Applications. Prentice-Hall, Englewood, N.J.
HERRICK, C. N. 1975. Electric Wiring Principles and Practices. Prentice-Hall, Englewood, N.J.
HUBERT, C. I. 1961. Operational Electricity. John Wiley and Sons, N.Y.
KLAYTON, M. H. 1977. Fundamental Electrical Technology. Addison-Wesley Publishing, Reading, Mass.
SCHICK, K. H. 1975. Introduction to Electricity. McGraw Hill, N.Y.

2

Resistive Networks

A first step in analyzing an electrical device or network is to describe its equivalent circuit in terms of ideal electrical elements. After completing the description, all the currents and voltages can be determined by applying three fundamental laws of circuits: Ohm's Law, Krichoff's Voltage Law, and Kirchoff's Current Law. These laws will be used to solve series, parallel and combination series — parallel circuits.

2.1 CIRCUITS AND CIRCUIT ELEMENTS

A circuit is defined as a continuous conducting path. A completely closed path must be present before a circuit exists. Circuits are made up of elements classified in two categories. Active elements are sources of energy, while passive elements are those elements which are not energy sources.

One type of active element is a voltage source. An ideal voltage source produces any amount of current while maintaining a constant voltage. An ideal voltage source has no internal resistance. A second active element is a current source. An ideal current source will produce any voltage across its terminals necessary to maintain a constant current output. It has infinite internal resistance. Figure 2.1 shows the symbols used for ideal current and voltage sources.

Resistors, capacitors and inductors make up the passive elements in circuits. Only resistors will be considered in this chapter. Capacitors and inductors will be introduced in a later chapter. Resistors are elements which place a specified amount of resistance in a circuit.

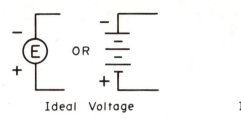

Ideal Voltage

Ideal Current

FIG. 2.1. IDEAL SOURCE ELEMENTS

Resistors can be constructed of materials such as carbon, metals and semiconductors.

Circuit elements will be connected by ideal conductors. Ideal conductors allow the current to flow with no voltage drop (no resistance in conductor).

2.2 SERIES AND PARALLEL NETWORKS

The simplest forms of network configurations are those in which resistive elements and all sources are joined end to end to form a single loop (*series* circuit) or all are connected between a single pair of terminals (*parallel* circuit).

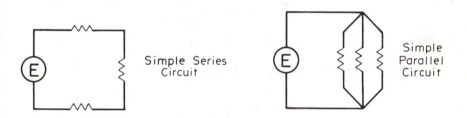

Simple Series Circuit

Simple Parallel Circuit

In the next section, we will learn how to apply Ohm's and Kirchoff's Laws to these circuits. This will allow us to solve for current flow and voltage drop for each element of a circuit.

2.3 SERIES CIRCUIT RELATIONSHIPS

A simple series circuit is made up of an ideal source, ideal resistors in which electrical energy is expended (commonly referred to as loads) and ideal conductors to connect the elements in series. A simple series circuit will have only one path for current to follow. The circuit in Example 2.1 will be used to demonstrate series circuit calculation.

Example 2.1.

Simple Series Circuit.

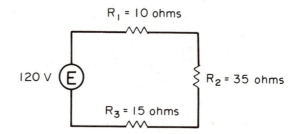

Kirchoff's Voltage Law states, the algebraic sum of all the voltage drops around any closed loop is equal to zero. For most applications this implies for a series circuit the sum of the voltage drops of the load elements (resistors) must equal the voltage of the source, i.e.,

$$E_S = E_1 + E_2 + E_3$$

where E_S = voltage of source

E_1 = voltage drop across resistor 1

E_2 = voltage drop across resistor 2

E_3 = voltage drop across resistor 3

There is only one path for current flow in a series circuit. Therefore, the current (electron flow) must be the same in each element of the circuit, i.e.,

$$I_S = I_1 = I_2 = I_3 = I$$

It is often necessary to determine the equivalent or total resistance of a number of resistors in series. We can develop an expression for total resistance by considering the fact that for series circuits, current in each resistor is equal and the sum of the voltage drops must equal the source. For a series of n resistors, the total voltage expression is

$$E_S = E_1 + E_2 + E_3 + \ldots + E_n$$

Using Ohm's Law for each resistor

$$E_1 = IR_1 \qquad E_2 = IR_2 \ldots \text{ and } E_S = IR_T$$

and substituting into the voltage relation we obtain

$$IR_T = IR_1 + IR_2 + IR_3 + \ldots + IR_n$$

which reduces to

$$R_T = R_1 + R_2 + R_3 + \ldots + R_n$$

The expression says the combined resistance of various loads in series is the sum of the separate loads.

In summary, there are three rules governing the simple series circuits with resistive elements. Demonstrated for Example 2.1, they are:

1. The current is the same in each element of the series circuit

$$I_S = I_1 = I_2 = I_3$$

2. The combined resistance of the various loads in series is the sum of the separate resistances.

$$R_T = R_1 + R_2 + R_3$$

3. The voltage across the source (power supply) is equal to the sum of the voltage drops across the separate loads in series.

$$E_S = E_1 + E_2 + E_3$$

Solution for Example 2.1.

We can now determine the current flow and voltage drop for each element of the circuit.

Recalling that Ohm's Law can be applied to the whole circuit or any part of the circuit, we can determine the current flow in the circuit. First calculating the total resistance using Rule (2),

$$R_T = R_1 \quad + R_2 \quad + R_3$$

$$= 10 \text{ ohms} + 35 \text{ ohms} + 15 \text{ ohms}$$

$$= 60 \text{ ohms}$$

Applying Ohm's Law, we find

$$I_S = \frac{E_S}{R_T} = \frac{120 \text{ V}}{60 \text{ ohms}} = 2 \text{ amps}$$

We can find the voltage drop for each resistor by again applying Ohm's Law, this time to each of the resistors.

Recall from Rule (1): $I_S = I_1 = I_2 = I_3$

$$E_1 = I_1 R_1 \qquad\qquad E_2 = I_2 R_2$$

$$= (2 \text{ amp}) (10 \text{ ohms}) \qquad = (2 \text{ amp}) (35 \text{ ohms})$$

$$= 20 \text{ V} \qquad\qquad = 70 \text{ V}$$

$$E_3 = I_3 R_3$$

$$= (2 \text{ amp}) (15 \text{ ohms})$$

$$= 30 \text{ V}$$

Note we can check our calculations by using the third rule:

$$E_S = E_1 + E_2 + E_3$$

$$120 \text{ V} = 20 \text{ V} + 70 \text{ V} + 30 \text{ V}$$

2.4 PARALLEL CIRCUIT RELATIONSHIPS

A simple parallel circuit is made up of an ideal source and ideal resistors connected in parallel across the source. The circuit in Example 2.2 demonstrates two ways of drawing the same circuit.

Example 2.2.

Simple Parallel Circuit.

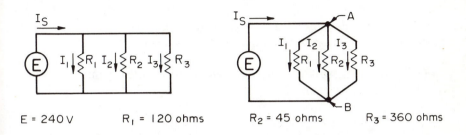

E = 240 V   R₁ = 120 ohms   R₂ = 45 ohms   R₃ = 360 ohms

In the parallel system, each of the load elements is connected across the same terminals, therefore each has the same voltage drop, i.e.,

$$E_S = E_1 = E_2 = E_3 = \ldots .$$

The total current from the source, however, divides at one terminal and recombines to return to the source at the second terminal. For the example, the current divides at point A, part of the current flowing through each resistor. The current then recombines at point B.

The second law, Kirchoff's Current Law, states that the sum of the current leaving a junction must equal the sum of the currents entering the junction. This can be expressed for the example as,

$$I_S = I_1 + I_2 + I_3$$

The expression for combined resistance of parallel elements can now be developed using the above expression and applying Ohm's Law to each element, i.e.,

$$I_1 = \frac{E_1}{R_1}, \; I_2 = \frac{E_2}{R_2} \ldots \text{ and } I_S = \frac{E_S}{R_T}$$

where

$$E_1 = E_2 = E_3 = E_S$$

Therefore

$$\frac{E}{R_T} = \frac{E}{R_1} + \frac{E}{R_2} + \frac{E}{R_3} + \ldots + \frac{E}{R_n}$$

The expression for total resistance becomes

$$\frac{1}{R_T} = \frac{1}{R_1} + \frac{1}{R_2} + \frac{1}{R_3} + \ldots + \frac{1}{R_n}$$

In summary, there are three rules governing simple parallel circuits with resistive elements. Demonstrated for Example 2.2, they are:

(1) Voltage across the parallel combination is the same as the voltage across each branch.

$$E_S = E_1 = E_2 = E_3$$

(2) The current flowing through the parallel combination (source current) is the sum of the currents in the separate branches (Kirchoff's Current Law).

$$I_S = I_1 + I_2 + I_3$$

(3) Summing resistances of a parallel circuit can be stated as follows: THE RECIPROCAL OF THE TOTAL RESISTANCE IS EQUAL TO THE SUM OF THE RECIPROCALS OF EACH OF THE INDIVIDUAL RESISTANCES.

$$\frac{1}{R_T} = \frac{1}{R_1} + \frac{1}{R_2} + \frac{1}{R_3}$$

Solution for Example 2.2.

Applying Rule Three (3) to Example Circuit 2, we can find the total resistance of the circuit to be

$$\frac{1}{R_T} = \frac{1}{120 \text{ ohms}} + \frac{1}{45 \text{ ohms}} + \frac{1}{360 \text{ ohms}} = \frac{(3 + 8 + 1)}{360 \text{ ohms}}$$

$$R_T = 30 \text{ ohms}$$

We can now determine the source current by applying Ohm's Law for the total circuit.

$$I = \frac{E_S}{R_T} = \frac{240 \text{ V}}{30 \text{ ohms}} = 8 \text{ amp}$$

We can also determine the current flow in each resistor by applying Ohm's Law to each.

Recall from Rule (1) for parallel: $E_S = E_1 = E_2 = E_3$.

$$I_1 = \frac{E_1}{R_1} \qquad\qquad I_2 = \frac{E_2}{R_2} \qquad\qquad I_3 = \frac{E_3}{R_3}$$

$$= \frac{240\ V}{120\ ohms} \qquad = \frac{240\ V}{45\ ohms} \qquad = \frac{240\ V}{360\ ohms}$$

$$= 2\ amps \qquad\qquad = 5.33\ amp \qquad\qquad = 0.67\ amp$$

We can check our calculations by applying Rule Two (2)

$$I = I_1 + I_2 + I_3$$

$$8\ amp = 2\ amp + 5.33\ amp + 0.67\ amp$$

2.5 COMBINATION SERIES — PARALLEL

It is not possible to solve circuit problems which are combinations of elements in series and parallel by direct application of our basic rules. However, voltage drop and currents in many combination series-parallel circuits can be determined by using basic rules to reduce the circuit to an equivalent simple series or simple parallel circuit. After one or more applications of the rules for addition of resistances in series or parallel, values in an equivalent simple circuit can be solved. Results can then be used to solve for values in the more complex circuit. The following example will demonstrate this method of solution.

Example 2.3.

Combination Series — Parallel Problem.

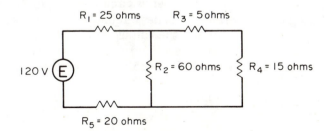

Step (1): Reduce by combining resistor in series or parallel to equivalent, thereby producing an equivalent circuit which can easily be solved. For the example shown, two combinations will be required.
(1a) Combining R_3 and R_4 (resistors in series) to R_{e34}

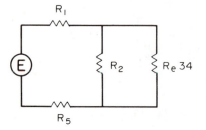

$$R_{e34} = R_3 + R_4 = 5 \text{ ohms} + 15 \text{ ohms} = 20 \text{ ohms}$$

(1b) Combine R_2 and R_{e34} (resistors in parallel) to R_{e234}

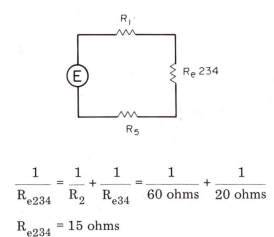

$$\frac{1}{R_{e234}} = \frac{1}{R_2} + \frac{1}{R_{e34}} = \frac{1}{60 \text{ ohms}} + \frac{1}{20 \text{ ohms}}$$

$$R_{e234} = 15 \text{ ohms}$$

Step (2): Solution of the simplified circuit by application of the basic rules. We now have a simple series circuit for our example problem. Applying the basic rules for series circuits and using Ohm's Law, we can solve for the current and the voltage drop across each element.

$$R_T = R_1 + R_{e234} + R_5 = 25 \text{ ohms} + 15 \text{ ohms}$$

$$+ 20 \text{ ohms} = 60 \text{ ohms} \qquad \text{(Series Rule (2))}$$

Applying Ohm's Law for the total circuit,

$$I_S = \frac{E_S}{R_{Total}} = \frac{120 \text{ volts}}{60 \text{ ohms}} = 2 \text{ amp}$$

Recall for a simple series circuit:

$$I_S = I_1 = I_{e234} = I_5 \qquad \text{(Series Rule (1))}$$

Applying Ohm's Law to each element,

$$E_1 = I_1 R_1 \qquad\qquad E_{e234} = I_{e234} R_{e234}$$

$$= (2 \text{ amp}) (25 \text{ ohms}) \qquad = (2 \text{ amp}) (15 \text{ ohms})$$

$$= 50 \text{ V} \qquad\qquad = 30 \text{ V}$$

$$E_5 = I_5 R_5$$

$$= (2 \text{ amp}) (20 \text{ ohms})$$

$$= 40 \text{ V}$$

Step (3): Using the information from the equivalent circuit, work back towards the original circuit. This will again require two steps for the example problem.

(3a) Knowing the voltage drop across $R_{e234} = 30$ volts, we see the voltage across the two parallel resistors R_2 and R_{e34} must also be 30 volts.

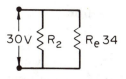

Therefore we can now solve for the current flow in each of the resistors.

$$I_2 = \frac{E_2}{R_2} = \frac{30 \text{ V}}{60 \text{ ohms}} = 0.5 \text{ amp}$$

$$I_{e34} = \frac{E_{e34}}{R_{e34}} = \frac{30\ V}{20\ ohms} = 1.5\ amp$$

(Note: $I_{e234} = I_2 + I_{e34}$; 2 amp = 0.5 amp + 1.5 amp)

(3b) Knowing the current flow through R_{e34} = 1.5 amp, we now know the current flow through each of the two resistors in series must be 1.5 amp. Therefore we can solve for the voltage drops across resistors R_3 and R_4.

$$E_3 = I_3 R_3 \qquad\qquad E_4 = I_4 R_4$$

$$= (1.5\ amp)\ (5\ ohms) \qquad = (1.5\ amp)\ (15\ ohms)$$

$$= 7.5\ V \qquad\qquad = 22.5\ V$$

We have now successfully solved for the current flow through, and the voltage drop across, each element of the series-parallel combination circuit.

EXERCISES

1. Draw a diagram, for the three resistors given below, if the resistances are connected to a 100 V source
   a. with all resistors in parallel,
   b. with all resistors in series.

   $$R_1 = 20\ ohms \qquad R_2 = 50\ ohms \qquad R_3 = 30\ ohms$$

2. a. What is the current flow through each of the resistors for problem 1a?
   b. What is the voltage drop across each of the resistors for problem 1b?
3. Determine the voltage drop across, and the current flow through, each of the resistors in the following circuits. At the end of each problem, set up a table of resistor number, current, and voltage.

(a.)

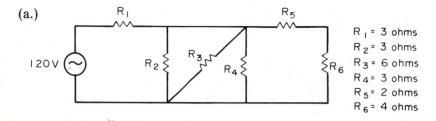

R_1 = 3 ohms
R_2 = 3 ohms
R_3 = 6 ohms
R_4 = 3 ohms
R_5 = 2 ohms
R_6 = 4 ohms

(b.)

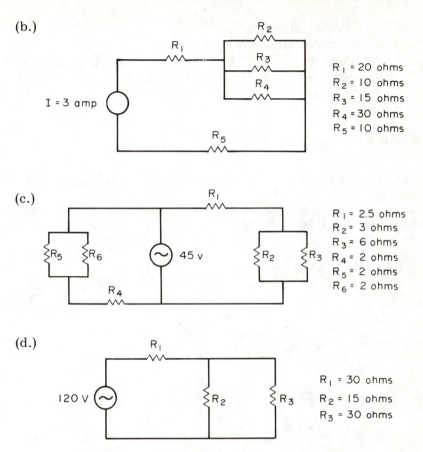

R_1 = 20 ohms
R_2 = 10 ohms
R_3 = 15 ohms
R_4 = 30 ohms
R_5 = 10 ohms

I = 3 amp

(c.)

R_1 = 2.5 ohms
R_2 = 3 ohms
R_3 = 6 ohms
R_4 = 2 ohms
R_5 = 2 ohms
R_6 = 2 ohms

45 v

(d.)

R_1 = 30 ohms
R_2 = 15 ohms
R_3 = 30 ohms

120 V

4. When building a voltmeter, a resistor is placed in series with the movement coil. If the movement coil of a DC voltmeter has a resistance of 30 ohms and gives a full scale reading with a current of 0.01 amperes in the coil, what size resistor is needed in order to have the meter read full-scale when measuring 100 volts?

5. Lighting load (120V) in a building was found to include:

    6 - 75 W lights
    10 - 100 W lights
    12 - 150 W lights
    6 - 300 W lights.

Remembering that the bulbs are all in simple parallel, calculate:

(a) resistance of each type of bulb,
(b) current flow in each type of bulb,

(c) total load in kilowatts,
(d) source current to supply all the lights.
(e) Assuming that 70% of the input to the lamps is given off as heat, what would be the amount of heat in Joules produced in one hour.

REFERENCES

MERKEL, J. A. 1974. Basic Engineering Principles. AVI Publishing, Westport, Conn.
WILCOX, G. and HESSELBERTH, C. A. 1970. Electricity for Engineering Technology. Allyn & Bacon, Boston.

3

Capacitance, Inductance and Phase Relations

Most real circuits contain loads which are not purely resistive. These loads will add an element of capacitance or inductance to the circuit. Resistive, capacitive, and inductive elements make up the three basic types of elements in all circuits. The following chapter deals with inductance and capacitance and how they effect the circuit.

3.1 INDUCTANCE AND INDUCTIVE REACTANCE

A coil of wire, such as in a transformer winding, is an important part of many pieces of electrical equipment. The coil is sometimes called an *inductor*. The following examples indicate the various uses of an inductor (coil) in electrical applications.

A. A relay, commonly used in switching circuits, has an iron core inside a coil of wire. When an AC current is applied to the coil it acts as an electromagnet closing or opening the contact points.

B. A transformer operates on the principle that an alternating current in one coil of wire will induce a voltage in a second coil. Transformers will be studied in more detail.

C. Motor windings are used to produce magnetic fields or to induce currents that allow development of mechanical force from electrical energy.

It is obvious that the inductor is an important electrical device.

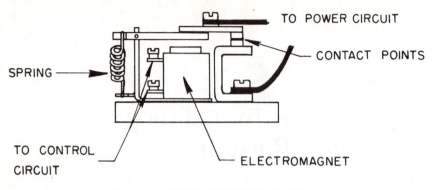

FIG. 3.1.   SCHEMATIC SKETCH OF RELAY

Note that each of the examples given made use of alternating current. The coil has the property that it opposes any change in current. This property is called *inductance*. The amount of opposition to current change is called *inductive reactance* and is a function of the frequency of the source and the inductance. Because inductance is dependent on the change in current flow, its primary applications are with alternating current where current varies cyclically with time.

The effect of inductance in alternating current circuits comes from changes in magnetic fields around a conductor. Every conductor with current flowing through it has a magnetic field around it. This magnetic field is a form of energy. Two important things to recognize about the field around a conductor are: first — the strength of the magnetic field is directly proportional to the magnitude of current flowing in the conductor; and second — the polarity of the magnetic field, north and south, is determined by the direction of current flow.

Consider the AC current wave shown in Fig. 3.2. At zero current, there is no magnetic field. As the current gets stronger the strength of the magentic field around the wires increases and more energy is stored. This continues until the current reaches a maximum at the top of the sine wave.

As the current decreases, energy stored in the magnetic field is returned to the circuit. When the current goes through zero and reverses direction the polarity of the magnetic field must reverse. The important factor is that creating the magnetic field, and thereby storing energy, requires time, as does the return of the energy when the field is reduced. This time requirement acts electrically the same as inertia acts on physical components. Electrically, the inertia of

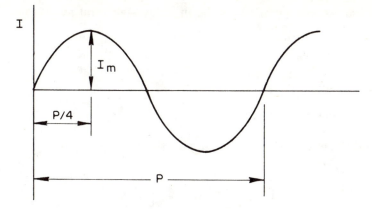

FIG. 3.2.   AC CURRENT WAVE

magnetic field causes the current to lag behind the voltage that causes the current to flow. The inertia or inductance creates an apparent counter-voltage or counter-emf which opposes the source voltage.

Inductance is measured in units of henries. A circuit or coil has an inductance of one henry when current varying at the rate of one ampere per second induces a counter-emf of one volt across the terminals of the coil.

The opposition to current flow in an inductor is called inductive reactance and is measured in units of ohms. The following formula is used to calculate the inductive reactance,

$$X_L = 2\pi fL \text{ ohms}$$

where   f = frequency in hertz, Hz

L = inductance in henries, H

π = 3.1416

Example 3.1.

Inductive Reactance.

If a coil known to have an inductance of 0.1 henries is connected to a 60 hertz source, what is the inductive reactance?

$$X_L = 2\pi \times 60 \text{ H}_Z \times 0.1 \text{ H} = 37.7 \text{ ohms}$$

The coil has an inductive reactance of 37.7 ohms.

We can express Ohm's Law for an inductive circuit as

$$E = I X_L$$

Example 3.2.

Inductance Problem.

Find the current flow in the circuit shown.

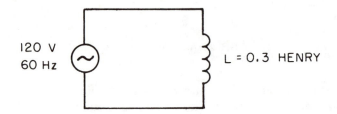

Finding the inductive reactance

$$X_L = 2\pi fL$$

$$= (2\pi)(60\text{Hz})(0.3\text{H})$$

$$= 113.1 \text{ ohms}$$

Applying Ohm's Law

$$I = \frac{E}{X_L} = \frac{120 \text{ V}}{113.1 \text{ ohms}}$$

$$= 1.06 \text{ amps}$$

Current in an AC circuit with pure inductance (negligible resistance) will lag the voltage by 90°. The figure below shows the current, voltage, and power wave forms for a pure inductance circuit. Note that the total power output of the circuit for any complete cycle is zero. This is also apparent from the AC power formula $P = EI\cos \phi$, where for this case the phase angle, $\phi = 90°$, therefore $\cos \phi = 0$ and true power output equals zero.

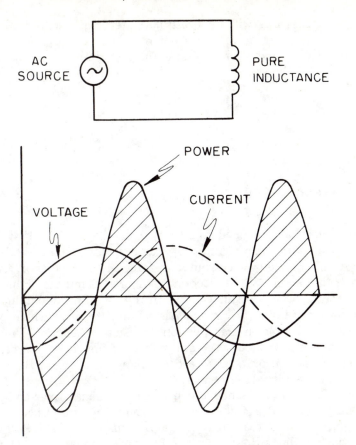

FIG. 3.3. PHASE AND POWER RELATION FOR INDUCTIVE CIRCUIT

We can express the voltage and current for a pure inductance circuit in phasor form as

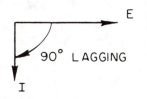

FIG. 3.4. PHASOR DIAGRAM FOR PURE INDUCTANCE

The phasor form will be useful in later work with impedance matching for power factor improvement.

Inductance of inductors in series and in parallel can be added using the following rules.

$$\text{Parallel} \quad \frac{1}{L_T} = \frac{1}{L_1} + \frac{1}{L_2} + \frac{1}{L_3} + \ldots$$

$$\text{Series} \quad L_T = L_1 + L_2 + L_3 + \ldots$$

3.2 TRANSFORMERS

Inductance can be divided into two categories. The first is self inductance. If the varying lines of magnetic force established by an AC current induce an electromotive force in the coil itself, the coil has self inductance. The second category is mutual inductance. Voltage transformers are an example of the use of mutual inductance.

The ordinary power transformer is a familiar object which plays a very important role in transmission of electrical energy. The transformer is used to change the voltage level of the power source. More detailed consideration of the need to change voltages in AC circuits will be discussed in a later chapter.

Figure 3.5 shows a schematic of a simple transformer containing a primary coil and a secondary coil. The primary coil has an alternating current in it that creates a varying magnetic field. Part of the magnetic field links the primary coil to the secondary coil. The varying magnetic field induces a voltage in the secondary coil. The iron core improves the inductive characteristics or the inductive transfer.

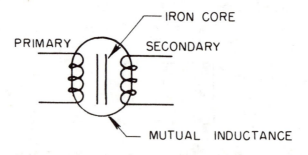

FIG. 3.5. SIMPLE TRANSFORMER

It is important to note that when a current flows and a magnetic field is produced in the primary, the induced current flow in the secondary will be in the opposite direction and will create a magnetic field of opposite polarity. This is stated in Lenz's LAW as "the induced voltage in a coil is always in a direction so as to oppose the effect which is producing it."

Power transformers are often used to change voltage levels. If the number of turns of wire in the primary and secondary coils is different, the voltage levels will vary as expressed by the equation

$$\frac{E_{primary}}{E_{secondary}} = \frac{N_{primary}}{N_{secondary}}$$

where   E = voltage

          N = number of turns of wire in coil

If the primary winding has more turns of wire than the secondary, the primary voltage will be higher than the secondary voltage. This would be termed a step-up transformer. Conversely, if the primary winding has fewer turns than the secondary, the secondary voltage will be lower than the primary. This is termed a step-down transformer.

Example 3.3.

Transformer Voltage Change.

What is the primary voltage of a transformer with a secondary voltage of 120 volts and a ratio of 20 primary turns to one secondary turn?

$$\frac{E_{primary}}{E_{secondary}} = \frac{N_{primary}}{N_{secondary}}$$

$$\frac{E_p}{120 \text{ V}} = \frac{20}{1} \qquad E_{primary} = 2400 \text{ V}$$

If there are an equal number of turns in both the primary and secondary windings, the induced voltage in the secondary is the same magnitude as the voltage applied to the primary. This type of transformer is called an isolation transformer and provides physical separation between a load circuit and the power source. Service men sometimes use the isolation transformer as a protective device against electric shocks and for the protection of expensive test equipment.

No power is lost in an ideal transformer, therefore, power in the primary circuit would equal power in the secondary circuit, i.e.,

$$P_{primary} = P_{secondary}$$

$$E_p I_p = E_s I_s$$

Using our formula for number of turns in each coil, we find

$$\frac{I_{secondary}}{I_{primary}} = \frac{N_{primary}}{N_{secondary}}$$

Example 3.4.

Transformer Current by Power Relation

A step-down transformer has a primary voltage of 120 volts and a secondary voltage of 24 volts. If a current of five amperes flows on the secondary side, what current is flowing on the primary side?

$$P_{primary} = P_{secondary}$$

$$E_p I_p = E_s I_s$$

$$120 \text{ V } I_p = 24 \text{ V} * 5 \text{ amp}$$

$$I_p = \frac{120}{120} \text{ amp} = 1 \text{ amp}$$

Example 3.5.

Transformer Current by Turns Relations.

If a transformer has a ratio of one primary turn to 10 secondary turns, what is the secondary current for a primary current of 20 amperes?

$$\frac{I_s}{I_p} = \frac{N_p}{N_s}$$

$$\frac{I_s}{20 \text{ amp}} = \frac{1}{10}$$

$$I_s = 2 \text{ amp}$$

Although transformers are very efficient devices, some losses do occur during transfer of power. They can be generally classed as:

(1) Copper losses (created by the resistance of the copper wire in the windings),
(2) Eddy-current loss (these can be overcome by making the transformer case out of many thin sheets of iron which are insulated from each other, called laminations),
(3) Hysteresis loss from molecular friction in the iron core (special silicon steel and heat treating processes for core materials are used to reduce this loss).

3.3 CAPACITANCE AND CAPACITATIVE REACTANCE

A capacitor is a basic electrical element found in controls, motors, welders, and many other places. A capacitor consists of two plates of electrical conducting material separated by an insulating material called the dielectric. Materials such as air, paper, mica, and oil can be used as dielectrics.

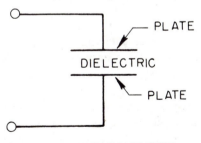

FIG. 3.6. SIMPLE CAPACITOR

When electric potential is connected to the plates an electric charge is stored in the capacitor. The plates of the capacitor will be charged as shown in the figure below when connected to a direct current source.

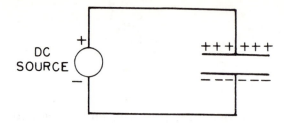

FIG. 3.7. CAPACITOR ON A DC SOURCE

Once the capacitor is charged, no current will flow in the circuit. A sample application of a capacitor in a DC system is a photo flashgun. Using the simplified schematic (Fig. 3.8), we see when the switch is open the battery will charge the capacitor. The capacitor stores the electrical energy. Action of the camera then closes the switch, thereby discharging the capacitor. Discharging the capacitor produces a large current flow through the light for a short period of time.

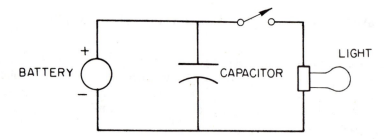

FIG. 3.8. FLASHGUN-DC APPLICATION OF A CAPACITOR

Connected to an AC source, a capacitor charges and discharges during each half-cycle (Fig. 3.9). A capacitor in an AC circuit does not block the current flow. Effect of the capacitor on the AC circuit will be discussed later in this section.

The amount of electric charge that a capacitor receives for each volt of applied potential is called its *capacitance*. Capacitance is measured in units of farads. The farad is a very large unit of capacitance. Practical devices are more often rated in terms of microfarads.

$$1 \text{ microfarad} = 1 \ \mu f = 10^{-6} \text{ farad} = \frac{1f}{1,000,000}$$

A capacitor in an AC circuit limits current flow in a similar manner to a resistor. The opposition to current flow is the *capacitive reac-*

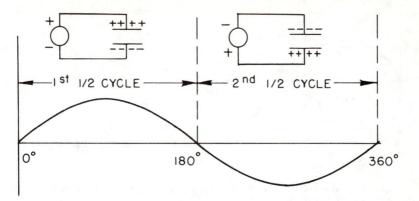

FIG. 3.9.   CAPACITOR ON AN AC SOURCE

tance. Capacitive reactance is measured in units of ohms and can be calculated from the frequency of the source and capacitance using the following formula:

$$X_C = \frac{10^6}{2\pi fC} \text{ohms}$$

where
        f = frequency in hertz, Hz
        C = capacitance in microfarads, μf

Example 3.6.

Capacitive Reactance.

Find the capacitive reactance of a 13 microfarads capacitor connected to a 60 hertz source.

$$X_C = \frac{10^6}{2\pi fC} = \frac{10^6}{(2\pi)(60 \text{ Hz})(13 \mu f)} = 204 \text{ ohms}$$

In an AC circuit with pure capacitance loading the current will lead the voltage by 90°, because of the time required to charge and discharge the capacitor. The diagram below shows the current leading the voltage in waveform and in phasor form.

Note that the phase angle is again 90°, which yields a power factor of zero and therefore a true power output of zero.

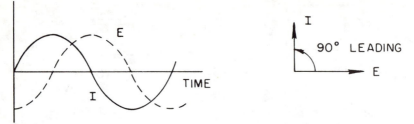

FIG. 3.10.   CURRENT AND VOLTAGE RELATIONS FOR PURE CAPACITANCE

Capacitance of capacitors in series and parallel can be added using the following rules:

$$\text{Parallel} \quad C_t = C_1 + C_2 + C_3 + \ldots$$

$$\text{Series} \quad \frac{1}{C_t} = \frac{1}{C_1} + \frac{1}{C_2} + \frac{1}{C_3} + \ldots$$

Ohm's Law can be applied in the form $E = IX_c$ for capacitors.

3.4 COMBINATIONS OF INDUCTANCE, CAPACITANCE, AND RESISTANCE

A generalized Ohm's Law is used to solve circuit problems which may contain all three types of elements: resistors, inductors, and capacitors. These circuits are sometimes called RLC circuits. The general form of Ohm's Law applied to RLC circuits states:

$$\text{Voltage} = \text{Current} \times \text{Total Impedance}$$
$$E = IZ$$

where
$$I = \text{current in amperes, amp, and}$$
$$Z = \text{total impedance, ohms.}$$

RLC circuits which are simple series circuits or simple parallel circuits will be studied in the following two sections. Combination series-parallel RLC circuits are beyond the scope of this text.

3.4.1 SIMPLE SERIES RLC CIRCUITS

It can be shown that for a simple series RLC circuit, as shown in Fig. 3.11, the total impedance is a vector sum of the resistance, capacitive reactance, and inductive reactance. Due to the phase shifts caused by capacitance and inductance, the impedances for a series circuit add vectorially as shown below:

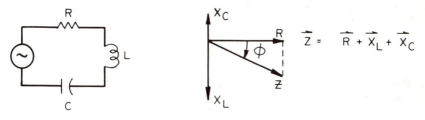

FIG. 3.11.   SIMPLE SERIES RLC CIRCUIT

The magnitude of the total impedance, Z, can be calculated as,

$$Z = \sqrt{R^2 + (X_C - X_L)^2} \text{ ohms}$$

The angle between the pure resistance, R, and the total impedance, Z, is the phase shift angle, ϕ. If Z is clockwise from R, the phase angle is considered leading; the current would be leading the voltage by the phase angle ϕ.

It can be shown that for a series RLC circuit the voltages of each element add vectorially in the same orientation as the impedances.

FIG. 3.12.   VECTOR ADDITION OF VOLTAGES FOR RLC SE-
RIES CIRCUIT

Again the magnitude of the sum of the voltages can be calculated as,

$$E_T = \sqrt{E_R^2 + (E_C - E_L)^2} \text{ volts}$$

Also the angle between the voltage due to pure resistance and the total voltage represents the phase shift angle, ϕ.

Using the generalized Ohm's Law and vector addition of impedances and voltages, problems involving series RLC circuits can be solved. Total impedance, current flow, phase shift angle, voltage drops for each element, and true and apparent power can all be determined.

The following series of examples help develop experience in working with series RLC circuits.

Example 3.7.

RC Series Circuit.

Solve for the current flow, power factor, and true and apparent power for the circuit below.

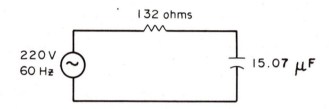

Determine the capacitive reactance.

$$X_C = \frac{10^6}{2\pi fC} = \frac{10^6}{(2\pi)\,(60\ \text{Hz})\,(15.07\ \mu f)} = 176\ \text{ohms}$$

Use vector addition to determine the total impedance and power factor.

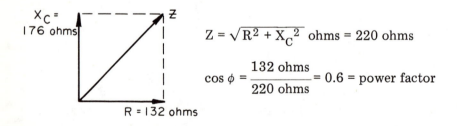

$$Z = \sqrt{R^2 + X_C^2}\ \text{ohms} = 220\ \text{ohms}$$

$$\cos\phi = \frac{132\ \text{ohms}}{220\ \text{ohms}} = 0.6 = \text{power factor}$$

Applying the general Ohm's Law,

$$I = \frac{E}{Z} = \frac{220\ \text{V}}{220\ \text{ohms}} = 1\ \text{amp}$$

True Power = $E I \cos\phi$ = 220 V $\times$ 1amp $\times$ 0.6 = 132 W

Apparent Power = EI = 220 V $\times$ 1 amp = 220 W

Example 3.8.

RL Circuit.

A current flow in the circuit of 10 amperes was measured. Determine the voltage drop across each element, the source voltage, the power factor, and the total impedance.

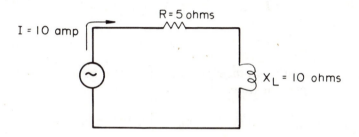

Ohm's Law can be applied to each element of the circuit to determine voltage drops:

$$E_R = I R = 10 \text{ amp} \times 5 \text{ ohms} = 50 \text{ V}$$

$$E_L = I X_L = 10 \text{ amp} \times 10 \text{ ohms} = 100 \text{ V}$$

The source voltage can be found by vector summation of the two voltages, E_L and E_R.

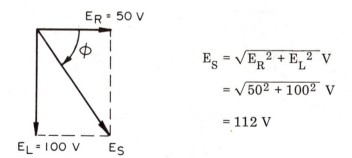

$$E_S = \sqrt{E_R{}^2 + E_L{}^2} \text{ V}$$

$$= \sqrt{50^2 + 100^2} \text{ V}$$

$$= 112 \text{ V}$$

Power Factor $= \cos \phi = \dfrac{50}{112} = 0.446$. Drawing the total impedance diagram,

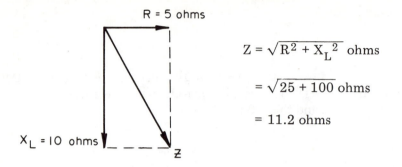

$$Z = \sqrt{R^2 + X_L^2} \text{ ohms}$$

$$= \sqrt{25 + 100} \text{ ohms}$$

$$= 11.2 \text{ ohms}$$

Note the source voltage can also be determined using total impedance and current flow.

$$E_{Source} = I \, Z = 10 \text{ amp} \times 11.2 \text{ ohms} = 112 \text{ V}$$

Example 3.9.

Series RLC Circuit.

Determine the total impedance, current flow, true power, and apparent power for the circuit shown.

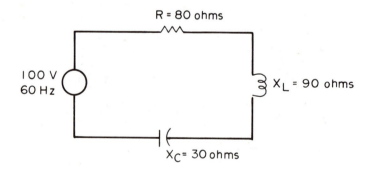

Obtained total impedance and power factor through vector addition.

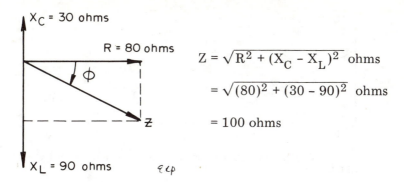

$$Z = \sqrt{R^2 + (X_C - X_L)^2} \text{ ohms}$$

$$= \sqrt{(80)^2 + (30 - 90)^2} \text{ ohms}$$

$$= 100 \text{ ohms}$$

$$\text{Cos } \phi = \frac{80}{100} = 0.8 = \text{Power Factor}$$

Determine the current flow.

$$I = \frac{E}{Z} = \frac{100 \text{ V}}{100 \text{ ohms}} = 1 \text{ amp}$$

True Power = E I Cos ϕ = 100 V $\times$ 1 amp $\times$ 0.8 = 80 W

Apparent Power = E I = 100 V $\times$ 1 amp = 100 W

3.4.2 Parallel Combinations of Inductance, Capacitance, and Resistance

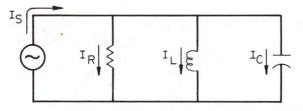

FIG. 3.13.   SIMPLE PARALLEL RLC CIRCUIT

When a circuit contains resistors, capacitors, and inductors in parallel, total current flow is obtained by vector addition of current flows for each element. Currents in this case add vectorially in the same orientation as the voltage in the series circuit, that is:

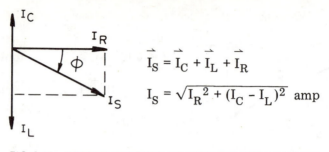

$$\vec{I}_S = \vec{I}_C + \vec{I}_L + \vec{I}_R$$

$$I_S = \sqrt{I_R^2 + (I_C - I_L)^2} \text{ amp}$$

FIG. 3.14. VECTOR ADDITION OF CURRENT FOR PARALLEL RLC CIRCUIT

The angle between the current flow due to pure resistance and the total current is the phase shift angle, ϕ, for the circuit.

Example 3.10.

Parallel Combination RLC.

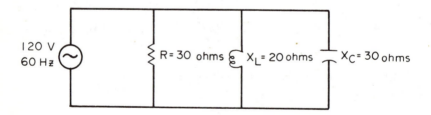

For the circuit shown, determine the current flow in each element, the source current, the true power, and the apparent power.

Use Ohm's Law for each element of the parallel combination. (Note the voltage across each element is the same for parallel connected loads.)

$$I_R = \frac{E_R}{R} \qquad\qquad I_L = \frac{E_L}{X_L} \qquad\qquad I_C = \frac{E_C}{X_C}$$

$$= \frac{120\text{ V}}{30\text{ ohms}} \qquad = \frac{120\text{ V}}{20\text{ ohms}} \qquad = \frac{120\text{ V}}{30\text{ ohms}}$$

$$= 4\text{ amp} \qquad\qquad = 6\text{ amp} \qquad\qquad = 4\text{ amp}$$

Determine the source current and phase angle by vector addition.

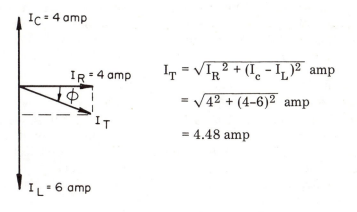

$$I_T = \sqrt{I_R{}^2 + (I_c - I_L)^2} \text{ amp}$$

$$= \sqrt{4^2 + (4\text{-}6)^2} \text{ amp}$$

$$= 4.48 \text{ amp}$$

$$\cos \phi = \frac{4}{4.48} = 0.89 = \text{Power Factor}$$

True Power $= EI \cos \phi$

$= 120 \text{ V} \times 4.48 \text{ amp} \times 0.89$

$= 478.5 \text{ W}$

Apparent Power $= EI = 120 \text{ V} \times 4.48 \text{ amp}$

$= 537.6 \text{ W}$

3.5 POWER-FACTOR IMPROVEMENT

The optimum situation for transmission of power exists when the power factor of the load equals unity; that is, when the phase shift between current and voltage is zero, $\phi = 0$. When the current is in phase with the voltage a minimum amount of current is required to deliver a given amount of true power. Minimizing the current flow minimizes the amount of power used to overcome the resistance of the transmission system.

Comparison of waveforms given earlier shows the power wave for a capacitive circuit differs in phase from the power wave of an inductive circuit by 180°. This implies that when the power wave of the capacitor is positive that of the inductor is negative, and vice versa. Because of the difference in phase the capacitor absorbs energy during the time the inductor releases energy and vice versa. The presence of both in the same circuit would result in the continuous alternating transfer of energy between the capacitor and the inductor thereby reducing the flow from the generator to the circuit. When balanced, all the energy released by the inductor is absorbed by the capacitor, and when released by the capacitor, it is absorbed by the inductor. Under this condition no reactive power is transmitted between the capacitor and the generator, nor between the inductor and the generator. Therefore, the generator supplies no reactive power to the circuit.

The nature of the circuit controls the choice of which type of element needs to be added to balance the circuit. That is, capacitive circuits require addition of inductance, and inductive circuits require addition of capacitance. In practice, the majority of electrical circuits contain motors, controllers, transformers, and other devices which create predominantly inductive loads. Under these conditions capacitors are needed to improve the power factor.

Power-factor improvement can be accomplished either by addition of series or parallel-connected capacitors. However, the parallel-connected capacitors have the advantage of not affecting the current through the other parallel-connected devices.

The following two examples demonstrate the procedure needed to calculate the required amount of capacitance.

Example 3.11.

Calculating Parallel-Connected Capacitance 1.

Given a 220 volt, single phase, 60 hertz inductive motor which draws 7.6 amperes with a power factor of 0.787, calculate the size of a parallel-connected capacitor required to return the power factor to unity.

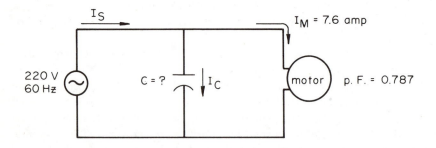

Solution:
  Phase Angle $= \phi = $ arc cos $(0.787) = 38°$
  Phasor diagram of motor current:

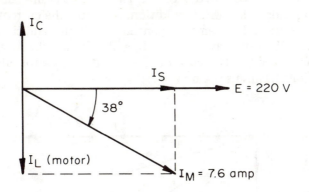

A current through the capacitor equal in magnitude to the inductive motor current, I_L, is required to balance the circuit.
Therefore,

$$I_C = I_L = 4.67 \text{ amp}$$

Using Ohm's Law to find capacitive reactance,

$$X_C = \frac{E}{I_C} = \frac{220 \text{ V}}{4.67 \text{ amp}} = 47.2 \text{ ohms}$$

Calculate capacitance necessary to obtain,

$$X_C = 47.2 \text{ ohms with a 60 Hz source.}$$

$$C = \frac{10^6}{2\pi f X_C} = \frac{10^6}{2\pi \times 60 \text{ Hz} \times 47.2 \text{ ohms}} = 56.3 \ \mu f$$

  Note the current through the motor is unaffected by the addition of the parallel connected capacitor.

Example 3.12.

  Calculating Parallel-Connected Capacitance 2.

Given the same motor as in Example 3.10, calculate the parallel-connected capacitance required to return the power factor to 0.95.

The phase angle of I_S after adding capacitance is known from the power factor, arc cos 0.95 = 18.2°. Therefore the line along which the new I_S falls is known. In addition, we know the current in-phase with the voltage, I_R, remains constant with the addition of pure capacitance. We can draw the new I_S and calculate the magnitude of new I_S and new I_L. The current through the capacitor added must equal the old I_L minus the new I_L.

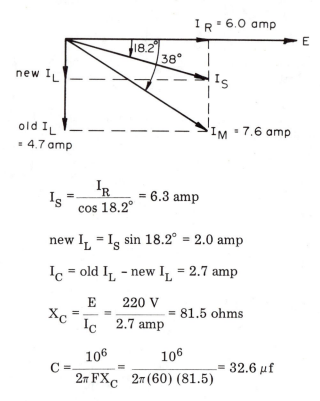

$$I_S = \frac{I_R}{\cos 18.2°} = 6.3 \text{ amp}$$

$$\text{new } I_L = I_S \sin 18.2° = 2.0 \text{ amp}$$

$$I_C = \text{old } I_L - \text{new } I_L = 2.7 \text{ amp}$$

$$X_C = \frac{E}{I_C} = \frac{220 \text{ V}}{2.7 \text{ amp}} = 81.5 \text{ ohms}$$

$$C = \frac{10^6}{2\pi F X_C} = \frac{10^6}{2\pi (60)(81.5)} = 32.6 \ \mu\text{f}$$

3.6 POWER-FACTOR IMPROVEMENT TABLE

A second method of determining size of capacitor necessary is by use of tables showing the relationship between the present power factor, load KVa, desired power factor, and the KVa of the rating of the parallel-connected capacitors needed to reach the desired power factor. This method will be demonstrated by the following example.

Example 3.13.

Use of Power Factor Correction Table A.8 of Appendix A.

Assume a 500 KVa load with a power factor of 0.6. It is desired to raise the power factor to 0.9. What size capacitor would be required for a parallel-connected capacitor bank? Use method of Table A.8. The capacitor rating in KVar is found by multiplying the true power consumption by the factor taken from the table. (Locate the original power factor of 0.60 in the first column on the left of the table and the desired power factor of 0.90 at the top of the table. Find the intersection of that row and column). The value of the factor for this example is 0.85.

$$\text{True Power} = \text{Apparent Power} \times \text{Power Factor}$$
$$= 500 \text{ KVa} \times 0.6$$
$$= 300 \text{ KW}$$
$$300 \text{ KW} \times 0.85 = 255 \text{ KVar}$$

Therefore we need a capacitor bank with a rating of 255 KVar.

Capacitance rating in KVar can be changed to rating of capacitor in farads if the voltage and frequency of the system are known.

Example 3.14.

Determining Capacitance from KVar.

Assume a voltage of 480 volts and a frequency of 60 hertz, what is the farad rating of a 255 KVar capacitor?
Calculate the capacitive reactance of the capacitor:

$$X_C = \frac{E^2}{P_C} = \frac{(480 \text{ V})^2}{255,000 \text{ var}} = 0.9035 \text{ ohms}$$

Calculate capacitance:

$$C = \frac{10^6}{2\pi f X_c} = \frac{10^6}{2\pi 60 \times 0.9035}$$

$$= 2.9 \times 10^3 \ \mu F$$

$$= 0.0029 \ f$$

3.7 ECONOMICS OF POWER-FACTOR CORRECTION

The operation of electrical systems at low power factors increases transmission costs. The higher current drawn by low power-factor loads causes greater line losses during transmission which then requires that large conductors be used. To encourage alleviation of such adverse conditions utilities companies may institute power-factor clauses in their rate structure. These clauses offer reduction in electrical rates for higher power-factor loads.

Addition of capacitors for power-factor improvement must be based on an economic analysis of the particular situation. The rate of return on capacitor investment will depend both on the structure of the power-factor clause and the present power factor. Power-factor improvements up to 90 to 95% are generally economically practical.

EXERCISES

1. For the circuit shown below,

    (a) Draw an impedance diagram.
    (b) Calculate the current through and voltage drop across each element.

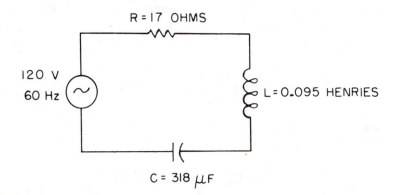

2. In the circuit shown, if E_R = 21 volts and E_L = 28 volts, find the values of

(a) E_S
(b) R
(c) X_L
(d) Z
(e) Power Factor
(f) True Power

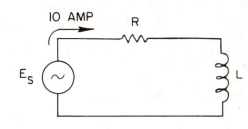

3. For the circuit shown, find the values of

(a) Current in each element,
(b) Total current,
(c) Power factor,
(d) True power,
(e) Total impedance.

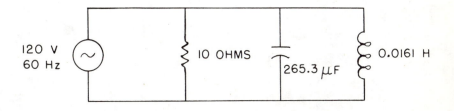

4. The following measurements were recorded for a prototype of an electric arc welder:

Voltmeter	Ammeter	Wattmeter
230 V	100 amp	13,800 W

What size capacitor would be required to improve the power factor to:

(a) 1.0,
(b) 0.90.

5. A small industrial plant has determined their load to be 50 KVa with a power factor of 0.76. What size parallel connected capacitor bank would be required to improve the power factor to 0.95?

REFERENCES

GERRISH, H. H. 1964. Electricity. Goodheart-Wilcox, Homewood, Ill.
HEIRICK, C. N. 1975. Electrical Wiring: Principles and Practices. Prentice-Hall, Englewood Cliffs, N.J.
HUBERT, C. I. 1961. Operational Electricity. John Wiley & Sons, N.Y.
KUBALA, T. S. 1974. Electricity 2. Delmar Publishers, Albany, N.Y.
MAPES, W. H. 1978. Power factor correction a la Mapes. Electrified Ind., April, p. 9-13 and May, p. 12-13.

4

Power Generation
and Distribution

In this chapter methods for production of electrical energy, commercial generation, transmission, and distribution of electricity, and basic concepts and devices necessary for safety in electrical systems will be discussed.

4.1 PRODUCTION OF ELECTRICAL ENERGY

There are a number of ways electrical energy can be produced. In all cases the source establishes a voltage which will cause a current to flow through a load when a complete circuit exists. Six primary methods of producing a voltage will be discussed in the following sections.

4.1.1 Friction

Voltage can be produced by rubbing two non-metallic materials together. Static electricity is often the undesired product of friction between moving objects. For example, the static electrical discharge experienced by crossing a wool rug in dry weather and then touching a metal object is static electricity. Clouds driven by strong winds can gather huge electrostatic charges. When released to earth in the form of lightning, they can do tremendous damage.

4.1.2 Pressure (Piezoelectricity)

Certain types of crystals produce a voltage when subjected to pressure. Microphone pickups, phonograph pickups, and accelerometers are example applications of piezoelectricity. Figure 4.1 shows a schematic of a vibration application, such as a microphone.

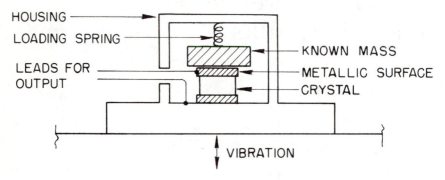

FIG. 4.1.   MICROPHONE PICKUP SCHEMATIC

4.1.3 Heat (Thermoelectricity)

A voltage will be produced when the junction of two unlike metals is heated. Thermocouples used for temperature measurement are primary examples of thermoelectricity. For the example thermocouple shown in Fig. 4.2, heat energy input to the hot junction forces the free electrons to move from the iron to the nickel. The resulting charges create a voltage of several millivolts (1/1000 of a volt) as long as the heat energy is input.

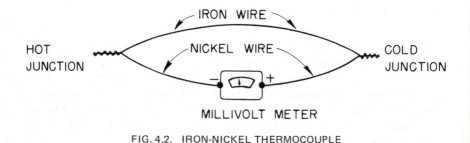

FIG. 4.2.   IRON-NICKEL THERMOCOUPLE

A number of thermocouples can be connected together to form a thermopile. Their combined output is used to operate such devices as safety valves, flame detectors, and thermometers.

4.1.4 Chemical Action

Chemical action can be used in two basic ways. The first is with batteries. Batteries rely on a chemical reaction to produce a voltage. Batteries are classified as primary cells if the chemical reaction is irreversible, which prohibits recharging the battery. If the chemical reaction is reversible, the battery is termed a secondary cell. The car battery shown schematically in Fig. 4.3 is an example of a secondary (rechargeable) cell.

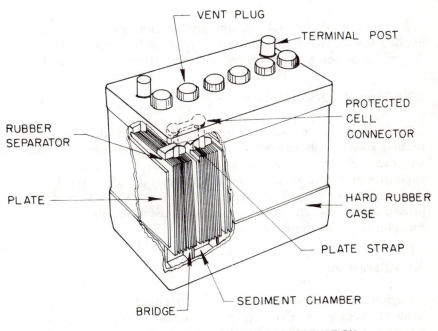

FIG. 4.3. 12 VOLT CAR BATTERY CONSTRUCTION

The second source of electrical energy through chemical action is the fuel cell. Devices which use up chemical elements or compounds to produce electrical energy are called fuel cells. Such cells have been used successfully in the U.S. space program. An example of a fuel

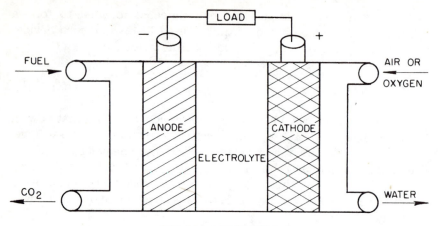

FIG. 4.4. A FUEL CELL

cell is shown in Fig. 4.4. Commonly, natural gas or liquid fuel is applied to one electrode while oxygen or air is applied to the other. Water and carbon dioxide are two common by-products of a fuel cell.

4.1.5 Light (Photoelectricity)

Devices which convert radiant energy to electrical energy are termed solar or photo cells. Light striking a photoelectric material will cause electrons to move in the material. Present solar cells are expensive and produce little energy per unit cost. However, considerable development work is under way which may lead to much improved solar cells. Photo cells will be discussed again when materials for solid state electronics are studied.

4.1.6 Magnetism

Magnetism produces a voltage by operating on the principle of induced voltage. In Fig. 4.5, a conductor is being moved rapidly downward through the magnetic field produced by the horseshoe magnet. As the conductor moves it cuts the lines of the magnetic field. This causes a voltage to be *induced* in the conductor. If a sensitive ammeter is connected across the ends of the conductor, the current created by the induced voltage will be indicated. It is evident that movement of the conductor through the magnetic field is responsible for the induced voltage since no current is present when the conductor is held motionless. The meter also indicates current

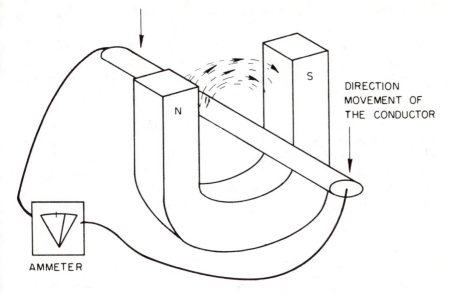

DIRECTION
MOVEMENT OF
THE CONDUCTOR

AMMETER

FIG. 4.5.   CREATING AN INDUCED VOLTAGE

flow in the opposite direction as the conductor is moved upward through the field. Therefore, polarity of the induced voltage is dependent on the direction of conductor movement through the field.

There are four factors which control the magnitude of the induced voltage:

(1)   the strength of the magnetic field,
(2)   the length of the conductor within the field,
(3)   the speed at which the conductor passes through the field,
(4)   and the angle at which the conductor passes through the field.

The following section will discuss the practical application of induced voltage for the production of electrical energy.

4.2 GENERATORS AND ALTERNATORS

Most electricity produced today is created through the use of alternators or generators. Both machines operate on the principle of induced voltage. The following sections will develop the operational principles of these machines.

4.2.1 Principles of Operation for an AC Generator

The essential components of a generator are shown in Fig. 4.6. A single conductor loop is placed so that it can be rotated in the space between two opposite poles of an electromagnet. To simplify the discussion one side of the loop is shown as shaded and the other side as white. In order to use the induced voltage in an external circuit each end of the loop is connected to these rings by means of a brush pressing against each ring. In other words, a complete circuit is provided through the sliding contacts at the slip rings.

The process within a generator as shown can best be demonstrated by following this loop as it moves through a complete revolution. Assume that the loop will be forced to move clockwise in

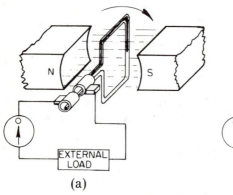
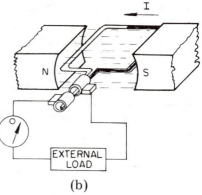

(a)   (b)

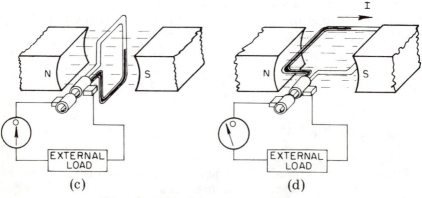

(c)   (d)

FIG. 4.6.   OPERATION OF AN AC GENERATOR

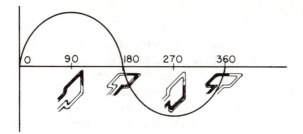

FIG. 4.7. INDUCED VOLTAGE FROM AN AC GENERATOR

the magnetic field. Figure 4.7 will be used to record the results of the induced voltage as the loop is rotated.

Starting in the position shown in Fig. 4.6 (a), the loop is moving parallel to the lines of the magnetic field. Therefore, because no lines of flux are being cut, no voltage is induced and no current flow will result in the load circuit.

As the loop rotates 1/4 turn to reach the position shown in Fig. 4.6 (b), the loop is cutting lines of the magnetic field at the maximum possible rate. At this point the induced voltage reaches it peak value. Current will also reach a peak value as it flows in the direction shown in the figure.

After rotating 1/2 turn to the position shown in Fig. 4.6 (c), the loop is again moving parallel to the lines of the magnetic field. At this point the induced voltage again drops to zero.

As the loop reaches the 3/4 turn position shown in Fig. 4.6 (d), the loop is again cutting a maximum number of lines of the magnetic field. However, at this point the loop is cutting the lines in the opposite orientation to the 1/4 turn position. Therefore, the induced voltage has a peak value of equal magnitude but opposite sign, and the current flow is in the opposite direction.

One quarter revolution later, the loop has reached its original position. At this point voltage and current will again be zero.

In summary, three important facts about the rotating loop should be emphasized:

(1) The induced voltage in the loop reverses in orientation twice each revolution.
(2) An alternating current present in the external circuit reverses its direction twice each revolution.
(3) The voltage and resulting current are pulsating in nature.

4.2.2 DC Generator

The single loop in a magnetic field can also be used to demonstrate the principles of a direct current generator. For the DC generator the slip rings will be modified to form the commutator. Figure 4.8 shows the single loop whose conductor terminates at a commutator consisting of a ring split lengthwise into two segments. Because the loop is rotating, a sliding contact is necessary to bring the current to the load circuit. Two brushes, connected to the load circuit leads, rest against the commutator segments.

Assume that the loop is again rotating in a clockwise direction. In position 1, Fig. 4.8 (a), the loop is not cutting the lines of the field. Therefore, no voltage or current is present. In position 2, Fig. 4.8 (b), the loop is cutting lines at a maximum rate. The induced

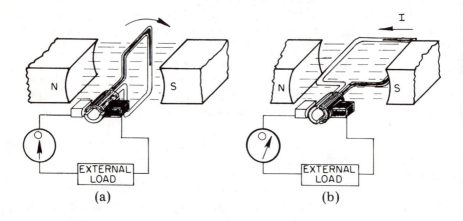

(a)   (b)

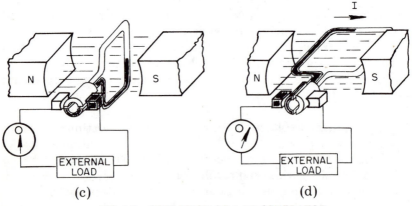

(c)   (d)

FIG. 4.8.   OPERATION OF A DC GENERATOR

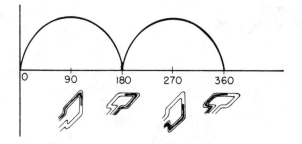

FIG. 4.9. INDUCTED VOLTAGE FROM A DC GENERATOR

voltage produces a current flow. In this position the black brush is the positive terminal and the white brush is the negative terminal of the generator. Note the current would be flowing from left to right through the load. In position 3, Fig. 4.8 (c), the loop is again in the vertical position. Therefore, no current or voltage exists. In position 4, Fig. 4.8 (d), note that the current in the loop flows in the opposite direction to that in position 2. However, because of the commutator, the current flow through the external load is in the same direction as before. Observe that the white brush is still negative and the black positive.

A graph of the voltage developed across the brushes, or as seen by the load, is shown for one revolution of the loop in Fig. 4.9.

In summary, three important facts about a DC generator are:

(1) The induced voltage in the loop reverses itself twice each revolution.
(2) The induced voltage and the resulting current are pulsating in nature.
(3) A pulsating direct current and voltage exist in the load circuit.

4.2.3 Basic Synchronous Alternator

Almost all commercial production of electrical energy is done with synchronous alternators. An alternator uses the same basic principles of induced voltage as the generator. However, in an alternator the magnetic field moves and the loops of wire remain stationary. Figure 4.10 shows the basic construction of a synchronous alternator. The field windings produce an electromagnet which is rotated about the center shaft of the alternator. A small DC generator, often on the same shaft as the windings, is used to power the electromagnet. The rotating magnetic field induces a voltage in the fixed windings.

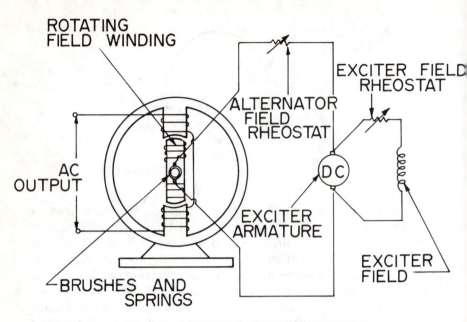

FIG. 4.10.   BASIC SYNCHRONOUS ALTERNATOR

There are several reasons for this reverse approach of alternators compared to generators. The rotating mass in the center is considerably lighter, requiring less concern about such design problems as the end bearing loads and braking. In addition, if very large generated voltages are taken off the brushes of the generator, heavy sparking and brush wear quickly result. With the alternator design the high voltages do not pass through the brush connectors because the wind-

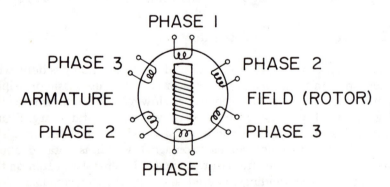

FIG. 4.11.   THREE-PHASE ALTERNATOR

ings are connected directly to the load. The DC voltages across the brush system of the alternator, to energize the electromagnet, are usually 240 volt or less.

Polyphase alternators can be built by adding more sets of windings. Figure 4.11 shows three sets of windings used in a three-phase alternator. The three sets of windings, spaced 120° from each other, produce three single-phase alternating outputs each with 120° phase difference from the other two.

The frequency of the sinusoidal voltage produced by AC generators and alternators is determined by the number of sets of windings used to produce each voltage and the speed at which the rotor turns, as shown in the following relation:

$$f = \frac{P\,N}{120} \text{(hertz or Hz)}$$

where   P = number of poles, (two per set of windings)
             N = speed of the rotor in RPM.

Example 4.1.

Determining Number of Poles for a Generator.

Determine the number of poles on the rotor of a single-phase generator, if the frequency of 60 hertz is generated at a rotor speed of 3600 RPM.

$$P = \frac{120\,f}{N} = \frac{120 \times 60}{3600} = \frac{7200}{3600} = 2 \text{ poles}$$

The frequency of each phase of a polyphase alternator or generator can also be calculated with the same formula.

4.3 POWER TRANSMISSION

4.3.1 Transmission System Concepts

To assist in understanding transmission systems, Fig. 4.12 demonstrates the basic components of an electric power system. The basic components consist of 1) the electric generating station, 2) the voltage increasing transformer or substation, 3) the high voltage

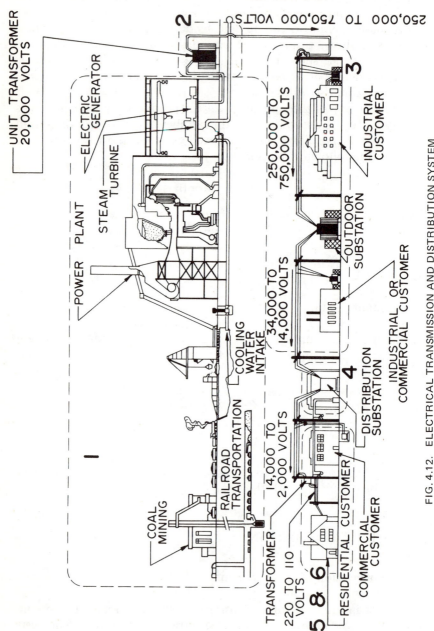

FIG. 4.12.   ELECTRICAL TRANSMISSION AND DISTRIBUTION SYSTEM

transmission lines, 4) the voltage decreasing transformer or substation, 5) the primary distribution network, and 6) the distribution service to the customer. Sections 2 through 4 are termed the transmission system, while 5 and 6 are the distribution system.

The transmission and distribution systems each contribute to carrying electric energy from generating facilities to the customer; however, the functions differ. The transmission system is designed to carry large quantities of energy over relatively long distances and may be referred to as the bulk system. The distribution system may be referred to as the local system and typically consists of several individual systems, each connected to one or more distribution substations. The distribution system carries energy over a network of low-voltage circuits to each customer.

4.3.2 High Voltage Transmission Lines

Electric transmission systems typically operate at relatively high voltages as compared to the voltages produced by the generator and those used in the distribution networks. This occurs because of the ability of a transmission line to more efficiently carry energy with increased voltage. Power lost in transmission (line loss as heat) is the product of the current carried and the resistance of the wire. If the resistance and current are held constant the line loss will remain constant. Increasing the voltage level allows the line to carry more power at the same current level. Therefore, as the voltage is increased the power transmitted is increased for the same line loss. For example, for the same current a 345,000 volt transmission line can carry three times the power of a 115,000 volt line with the same line loss.

4.3.3 Developing a Transmission System

A transmission system is to deliver the required amount of electric energy to the associated distribution systems with the necessary reliability, and at the lowest cost. It is not possible to design a system for minimum cost without recognizing the effects on the reliability of the system.

The concept of a transmission system can be understood by examining a very simple example. Figure 4.13 (a) shows a service area which is divided into four distribution networks. Distribution substations are shown as black dots. Distribution systems can economically transport energy only over relatively short distances, therefore several substations are necessary for the service area.

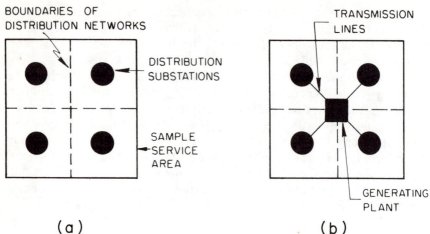

(a)   (b)

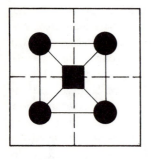

(c)

FIG. 4.13. ELECTRICAL TRANSMISSION AND DISTRIBUTION SYSTEM

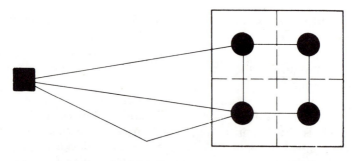

FIG. 4.14. TRANSMISSION SYSTEM WITH GENERATOR DISTANT FROM SERVICE AREA

In Fig. 4.13 (b), a generating station is located at the center of the area. A very simple system can be constructed by extending one transmission line from the generating station to each distribution substation. If the generating station is within the load region, the required transmission lines are relatively short. This system lacks reliability, however. Failure of a transmission line means energy can not be supplied to one part of the region.

To improve the reliability of the transmission system additional lines can be constructed between the distribution substations as indicated in Fig. 4.13 (c). In this system the loss of any one of the transmission lines from the generating station will result in increased energy flow over the remaining lines in such a manner that the required energy reaches each of the substations. However, the transmission lines must be designed with sufficient capacity such that shifting of the paths of energy flow does not overload and cause a failure of any link in the system. While the cost of this system is higher than that shown in Fig. 4.13 (b), the reliability is much more acceptable.

Figure 4.14 demonstrates the effect of locating a generating station a considerable distance from the area it serves. It is obvious here that not only does the system become more expensive but the greatly increased exposure of the system to natural hazards effectively reduces the reliability.

Thus far, the reliability of a system with only one generating plant has been discussed. Generators must routinely be removed for service and are also subject to failure. Therefore, it is necessary to link large areas together and to have some "spare" generating capacity in the overall region.

In summary, the function of a transmission system is to economically and reliably:

(a) transmit power from the generating stations to the load areas,
(b) interconnect load areas, generating stations, and individual systems to improve reliability,
(c) and interconnect utilities in a region to allow sharing of generation reserves and other benefits.

4.4 THREE-PHASE SYSTEMS

Where large quantities of electrical power are being transmitted or used three-phase AC power is generally used. As described in Section 4.2.3, such currents are generated by an alternator having three identical armature coils spaced $120°$ from each other. Since each coil has two connections, it at first appears that six lines are needed to transfer currents to the loads. Fortunately, it is possible to connect one end of each coil to another coil at the alternator and then transmit the current over three lines, one for each phase. Two basic configurations, called "wye" and "delta", are used. Each of these systems will be described in the following sections.

4.4.1 Three-Phase Delta

For the delta configuration, Fig. 4.15, the ends of each armature coil are joined to the ends of the other two coils to get the characteristic triangle, or delta (Δ), shape.

When the three-phase (3ϕ) lines are connected to the three-phase load as in Fig. 4.15, the voltage produced by a coil is called the phase voltage and the voltage between any two lines is called the line voltage. For a delta-connected circuit the phase voltage equals the line voltage.

The current through any one coil, when connected to the load, is called the phase current. The corresponding current flow through any line is called the line current. In a delta configuration a line cur-

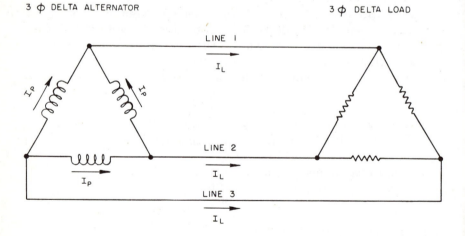

FIG. 4.15. THREE-PHASE DELTA CONFIGURATION

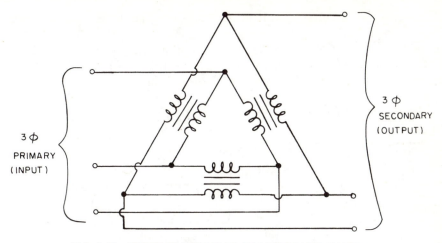

FIG. 4.16.   THREE-PHASE DELTA TRANSFORMER SYSTEM

rent is the vector sum of two phase currents. Because a phase difference of 120° exists between the two, it can be shown that the line
current equals $\sqrt{3}$ (or 1.73) times the phase current.

In Fig. 4.15 the loads on the system are assumed to be balanced.
That is, the load is such that all line currents are equal and have the
same phase angle with respect to line voltages. Power companies try
to maintain a balanced condition because it minimizes power losses.
A balanced load is assumed in all three phase systems discussed in
this book.

Step-up and step-down transformers are established by using three
transformers, identical to those used for single-phase systems, connected in the same delta configuration (Fig. 4.16).

To obtain single-phase AC from a delta system, connections are
made across any one phase of a transformer secondary. For example,
if the phase voltage of the secondary is 240 volts a single-phase 240
volt system can be obtained by connecting across one phase, as
shown across phase B in Fig. 4.17. If a single-phase 120/240 volt
system is needed, a neutral wire is connected to a center tap as
shown in phase C. In practically all installations the neutral wire is
grounded. Assuming a grounded neutral system, 240 volts are applied
to any device connected across the entire phase winding (between
lines 1 and 2), while 120 volts are delivered to any device connected
between the neutral and either of the outer (hot) wires (between 6
and 5 or between 4 and 5). More discussion of the 120/240 volt single-
phase system is included in a later section.

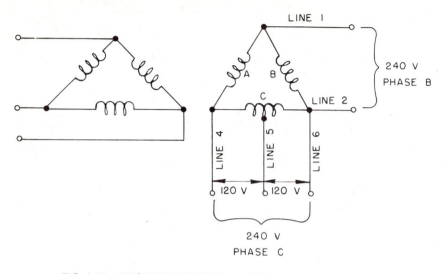

FIG. 4.17. SINGLE-PHASE FROM THREE-PHASE DELTA SYSTEM

4.4.2 Three-Phase Wye

The second configuration for three-phase systems is the Y-shaped arrangement where one end of each coil is connected to one end of each of the other coils (Fig. 4.18). The free ends are then the terminals for the three-phase lines. In the wye system configuration the line current now equals the phase current. However, the voltage between any two lines is now equal to the vector sum of two of the coils. For the wye system the line-to-line voltage equals the $\sqrt{3}$ (or 1.73) times the phase voltage.

As in the delta system, single phase may be obtained by connecting across any two lines.

A common variation of the wye configuration is known as the three-phase, four-wire system (Fig. 4.19). It has a grounded neutral connected to the common junction of the three transformers. This modified configuration offers advantages in available voltage levels. In the four-wire system, single-phase current can be obtained by connecting to the neutral and to any one of the other three hot wires. If, for example, the three-phase voltage is 208 volts the single-phase current will be 120 volts. Thus by using four wires we can obtain either a three-phase current at 208 volts for motors, water heaters, and similar large loads, a three-phase current at 230 volts for three-phase motors, or a single-phase current at 120 volts for lighting and small appliances.

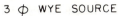

3 φ WYE SOURCE

3 φ LOAD

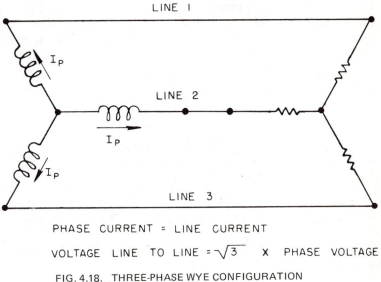

PHASE CURRENT = LINE CURRENT

VOLTAGE LINE TO LINE = $\sqrt{3}$ X   PHASE VOLTAGE

FIG. 4.18. THREE-PHASE WYE CONFIGURATION

    In summary, a prime reason for the use of three-phase systems is
that for the same voltage and current a three-wire, three-phase system will deliver 1.73 times as much power as a two-wire, single-
phase system. Thus with the addition of one more wire (50% more
wire), 73% more power can be transmitted. This helps minimize the
size and cost of transmission systems. In addition, the total power
output for a three-phase system is a constant value. As shown in

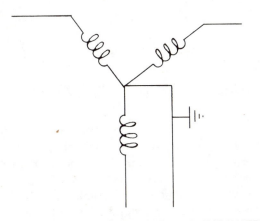

FIG. 4.19. THREE-PHASE, FOUR-WIRE WYE SYSTEM

Fig. 4.20, the power from each phase varies in a sinusoidal manner, as does a single-phase power curve. However, the total instantaneous power, which is the sum of the three curves, is constant over time. Total true power output for the three-phase system is expressed by:

$$P = 3E_p I_p \cos\phi$$

or

$$P = \sqrt{3}\ E_L I_L \cos\phi$$

where

E_p = phase voltage

I_p = phase voltage

E_L = line voltage

I_L = line current

$\cos\phi$ = phase shift angle.

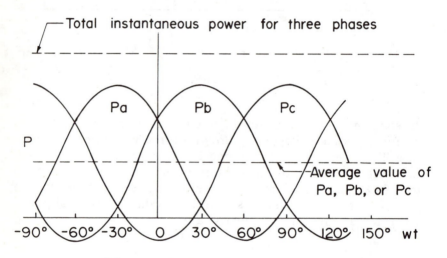

FIG. 4.20. POWER IN A THREE-PHASE SYSTEM

4.5 120/240 VOLT SINGLE-PHASE SERVICE

By far the most common service system for farms and residences is the 120/240 volt three-wire single-phase system (Fig. 4.21). Understanding how this system functions is important to planning and installing single-phase systems. This system will likely originate with a step-down transformer, where the primary side is the power supplier's distribution line and the secondary side is the customer's ser-

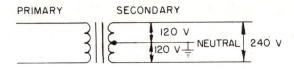

FIG. 4.21.   ORIGIN OF 120/240 VOLT SINGLE-PHASE SERVICE

vice drop. The transformer secondary will have a center tap where the neutral is connected. This neutral is also grounded and can therefore be considered at zero volts, relative to the earth. Between the neutral and either of the hot wires a potential of 120 volts AC will exist.

Consideration must be given to how loads are connected to this system and how the manner in which they are connected affects current level in each conductor. As shown in Fig. 4.22, if the neutral is established at zero volts potential, relative to the earth, we must have one hot conductor at a potential of +120 volts, relative to the neutral, and the second hot conductor at −120 volts. Note the potential difference between the two hot wires is 240V. Loads requiring a voltage potential of 240 volts are then connected across the two hot wires. Note that the neutral wire is not connected to these loads, therefore these loads place no current in the neutral wire.

Loads requiring a 120 volt potential may be connected between the neutral and either of the two hot wires. In this case current may be carried by the neutral wire. As will be demonstrated in the following example, the amount of current flowing in the neutral can be minimized by balancing the loads, that is, by having equal loads connected to each of the hot wires.

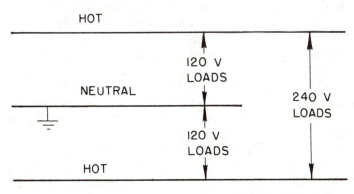

FIG. 4.22.   CONNECTING LOADS TO 120/240 VOLT SERVICE

Example 4.2.

Current Levels for 120/240 Volt System.

For the system shown schematically in the figure below, assume each of the 120 volt loads (A–D) draws 10 amperes when connected and the 240 volt load (E) is a motor drawing 20 amperes when connected. Determine the current flow in each conductor for the following cases:

(a) only load E connected
(b) only load A connected
(c) loads A and C connected
(d) loads A, B, C, and E connected

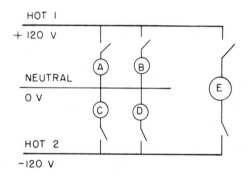

(a) If load E is connected, this load requires 20 amps current at 240 V. In this case we can establish that we must have a current flow of 20 amps in Hot 1, through the load, and out Hot 2, with no current flow in the neutral. Therefore, the currents in the three conductors, for this case, are: Hot 1, 20 amps; Neutral, zero amps; and Hot 2, 20 amps.

(b) If load A is connected, this load requires 10 amps. For this case the load would see a 120 V potential between Hot 1 and Neutral. Therefore, a current of 10 amps would flow in Hot 1 and out the Neutral conductor. Hot 1, 10 amps; Neutral, 10 amps; Hot 2, zero amps.

(c) If loads A and C are both connected, they will each require a current flow of 10 amps. As in case (b), 10 amps will flow in Hot 1 to supply load A. However, because load C is also connected, the current now sees a second path through C to Hot 2 with another voltage drop of 120 V. Therefore, the current will flow in Hot 1, through load A, through load C, and return through Hot 2. Because the loads on the two sides of the neutral are balanced no current will flow through the neutral. Hot 1, 10 amps; Neutral, zero amps; Hot 2, 10 amps.

(d) If loads A, B, C and E are connected, current flow will be required in the hot wires to supply both the 120 V loads and the 240 V load. Forty amps will be required to flow in Hot 1 to supply the loads connected to it (Load A, 10 amps; Load B, 10 amps; and Load E, 20 amps). However, only 30 amps will be returning through Hot 2 (Load C, 10 amps and Load E, 20 amps). The difference in current flows between the two hot wires (or the imbalance) will be carried by the neutral conductor. In this case a current of 10 amps will flow in the neutral conductor. Hot 1, 40 amps; Neutral, 10 amps; Hot 2, 30 amps.

When considering sizing wires for the 120/240 volt service, we can see that because the 240 volt loads do not impose any current on the neutral, and if the 120 volts are balanced between the two hot sides, current flow in the neutral wire will be considerably less than that in the hot wires. For this reason the National Electric Code commonly allows the neutral wire of such a system to be of a smaller size than the hot conductors.

4.6 PRINCIPLES AND PRACTICES FOR ELECTRICAL SAFETY

Principles of electrical safety can be divided into protection of the electrical system and equipment from damage due to overcurrents or overvoltages and protection of people and property from electrical shock or other harm. In this section we will consider several procedures and devices which are used in development of safe electrical systems.

4.6.1 Grounding

The term grounded, in electrical terminology, means connected, directly or indirectly, to the earth. The purpose of grounding is for safety. Our discussion of grounding will be divided into the categories of system grounding, which is grounding of current-carrying portions of the system, and equipment grounding, which is grounding of equipment not intended to be at a voltage potential different from the earth.

The principal reason for system grounding is to limit the voltage between any conductor and ground from rising above a safe level due to some fault outside the building. System grounding entails proper grounding of the service entrance neutral and connecting the neutral wire of all feeders and branch circuits to the grounded service neutral. The service entrance neutral is connected by a grounding-electrode conductor to a grounding electrode. This grounding electrode is normally a metal water pipe. Where a metal water pipe is not available other metal underground systems such as gas-piping systems can be used. A grounding electrode can also be constructed from a pipe, rod, or plate (see National Electric Code Section 250 H. for further specifications). System grounding protects from over-voltages if, for example, the outside wires are struck by lightning. The grounding acts like a lightning rod to dissipate large amounts of electrical charge.

Equipment grounding is necessary to prevent electric shock to persons coming into contact with metallic objects which, either intentionally or accidentally, form part of the electrical system. For example, if a faulty circuit or device exists such as an ungrounded (hot) conductor is touching any metallic equipment, junction box, motor frame, or fitting, then there is a voltage between that object and ground. If that object is not grounded a very serious shock hazard exists.

To illustrate the potential problem consider the situation in Fig. 4.23. If a fault develops in the motor windings such that the hot conductor comes in contact with the frame and the frame is ungrounded,

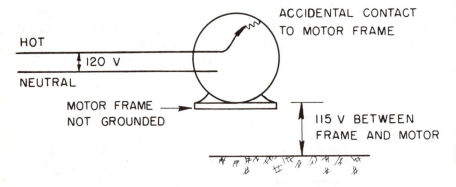

FIG. 4.23.   UNGROUNDED MOTOR WITH FAULT TO FRAME

a 120 volt potential exists between the motor frame and earth. If a person who is grounded by standing on the earth comes in contact with the frame he will be subjected to a very hazardous, if not fatal, shock. This type of hazard can be avoided by grounding the motor frame. If the hot conductor were to then come in contact with the frame, a short circuit would exist and the overcurrent protection for the circuit would then "blow" or open the circuit due to the high current flow.

The NEC requires that all metallic equipment — raceways, boxes, fittings, enclosures for fuses or circuit breakers, and so forth — is to be grounded. This grounding is accomplished by assuring that there is a low-resistance path established from all metal objects to ground. In many instances an extra conductor (green or uninsulated) will be used. It is important to note that the equipment grounding should never, under normal operation, carry any current. Only when a fault exists should current flow in the equipment grounding system. In addition, no switches or interruptions should exist anywhere in the grounding system.

4.6.2 Polarity and Switching

In most wiring systems, conventions have been established for the color of the covering of each conductor in the system. Using the established system is called polarizing and must be followed faithfully throughout the system. For example, in a single-phase system a white wire is always used as the neutral wire, red or black wires are used as the "hot" conductors, and green wire is used as the grounding wire. By connecting the white wires to silver-colored terminals and red or black wires to brass-colored terminals the circuit is polarized. If a circuit is correctly polarized, all switches will be placed in the hot (red or black) conductors. When a switch is connected in series with the grounded neutral conductor, as shown in Fig. 4.24 (a), a voltage potential exists between the electrical system of the device and the ground when the switch is open. If, for example, the device is a light bulb socket, the screw shell would be at a potential of 120 volts. This would make the changing of light bulbs hazardous. If the switch is located in series with the hot conductor, the object such as the light socket is a zero volts potential and presents no hazard to someone working around it when the switch is open. It should be apparent why *the grounded neutral is never to be interrupted by a fuse, circuit breaker, switch or other device.*

INCORRECT SWITCH LOCATION

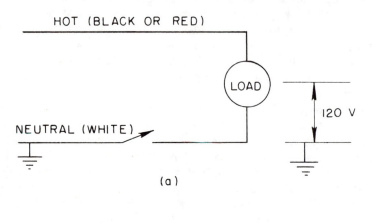

(a)

CORRECT SWITCH LOCATION

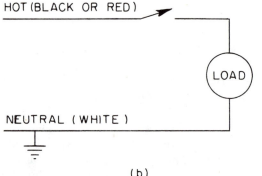

(b)

FIG. 4.24. CORRECT PLACEMENT OF SWITCH IN A SINGLE-PHASE SYSTEM

4.6.3 Overcurrent Protection

Any electrical system needs safeguards to assure that safe levels of current for conductors and equipment are not exceeded. Two basic classes of devices are used for this purpose, fuses and circuit breakers. Both are installed in series with the hot conductors and are designed to open the circuit if a specified current level is exceeded.

Fuses are the most common protective device. Fuses are overcurrent devices of which a portion is destroyed when interrupting the circuit. They are made with a low melting point metal link which is calibrated to melt when a specific current rating is reached. All fuses have an inverse time characteristic. Thus a standard fuse will carry a 10% overload for a few minutes before melting, a 20% overload for less than a minute, and a 100% overload for only fractions of a second. Often in circuits for control of motors, fuses which will allow a temporary overcurrent during motor starting are needed. Time-delay fuses are designed for this need. An example of a plug-type time-delay fuse is shown in Fig. 4.25.

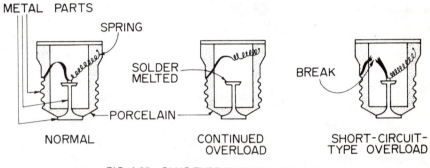

FIG. 4.25. PLUG-TYPE TIME-DELAY FUSE

A time-delay fuse can fail in two ways. During a continuous over-load, a solder connection at one end of the link will melt and the spring will pull the link away, breaking the circuit. During a short circuit, the link itself will melt almost instantaneously. However, an overload created by a motor start will not be large enough to melt the link or of long enough duration to melt the solder connection. This yields the time-delay protection.

When the circuit current exceeds 30 amperes, it is necessary to use a cartridge-type fuse and fuse holder. Cartridge fuses work on the same principle as plug fuses. A schematic of a time-delay cartridge fuse is shown in Fig. 4.26.

A circuit breaker is a device designed to open a circuit automatically on a predetermined overload current without damage to itself. Most circuit breakers have a bimetallic strip connected in series with the contacts (see Fig. 4.27).

Its function is to protect the circuit from a continuous overload. Current passes through the bimetallic strip causing it to heat up. As the element heats up, the two metals expand at different rates. This causes the element to bend. If the current level is too high, the bend

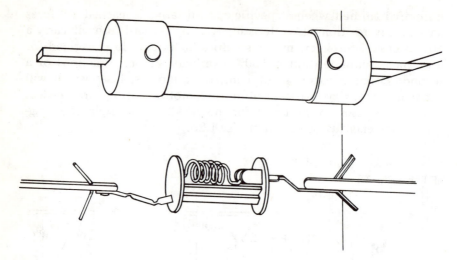

FIG. 4.26.   TIME-DELAY CARTRIDGE FUSE

will be large enough so that the contact points will be opened. After the element has cooled the circuit breaker can be reset.

Another type of circuit breaker contains an electromagnet. Normal current levels have no effect on the electromagnet. However, when a short circuit occurs the electromagnet trips the breaker immediately.

Combination thermal-magnetic circuit breakers are also available to protect against both continuous overloads and short circuits.

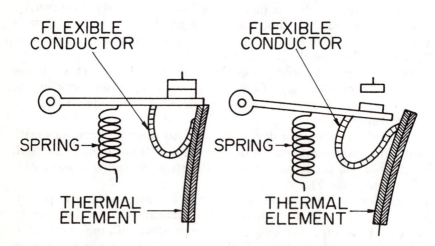

FIG. 4.27.   BASIC CONSTRUCTION OF A THERMAL CIRCUIT BREAKER

4.6.4 Ground Fault Circuit Interrupters (GFI)

In a complete electrical circuit there must be at least two wires, one to carry the current to the load and one to return the current to the outlet or source. If the insulation of the wiring or the load is faulty or breaks down, all or a portion of the current may follow an alternate path back. This situation is called a ground fault.

As demonstrated in Fig. 4.28, this fault current can be extremely hazardous if its path is through a person's body. A ground fault current may not be of sufficient magnitude to cause a fuse or circuit breaker to trip, but still may be fatal to a person. Figure 4.29 shows current levels at which a person can no longer let go of the device and at which fibrillation (loss of control of heart beat) occurs. These currents are much lower than those required to activate overcurrent protection.

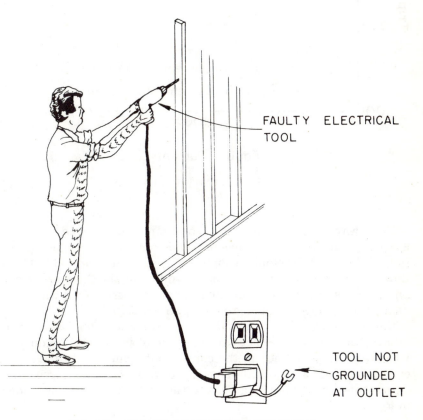

FAULTY ELECTRICAL
TOOL

TOOL NOT
GROUNDED
AT OUTLET

FIG. 4.28. GROUND FAULT FROM FAULTY TOOL

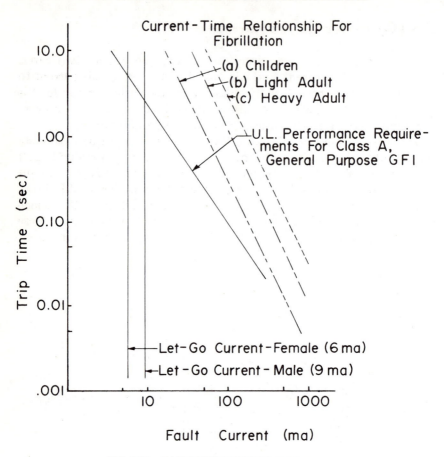

FIG. 4.29.   FAULT CURRENT RELATIONSHIPS

Three ways exist to protect a person from such a ground fault. One method is to use only double-insulated tools. This provides protection if the fault develops in the tool. The second method is to use three-wire (grounded) cords. This is typified by the three-prong plug on many appliances. All too often the third or grounding wire is not used, thereby rendering the grounding wire useless.

The third method is to use a device known as a ground fault circuit interrupter. This device measures the current in the two conductors. Whenever the difference in the current out compared to the return current exceeds a specified value the device opens the circuit. Requirements for opening by a Class A, General Purpose GFI are shown in Fig. 4.30.

Ground faults can occur almost anywhere but are most serious in wet or damp areas because moisture helps conduct the current. With this in mind, the National Electric Code requires GFI's in all circuits in new wiring supplying receptacles in the bathroom and garages of dwellings. They are also required on outdoor receptacles accessible from the ground level or associated with swimming pools.

It is important to note the difference in function between ground fault interrupters and overcurrent protection devices. Fuses and circuit breakers protect the system from excessively large currents while GFI's protect from leakages of currents.

GFI's are generally available in three forms. One type is a circuit breaker with built-in ground fault interruption, which can be used in place of standard circuit breakers. Another form is a GFI which can be used to replace a standard duplex receptacle. A third kind is a unit which plugs into standard receptacles.

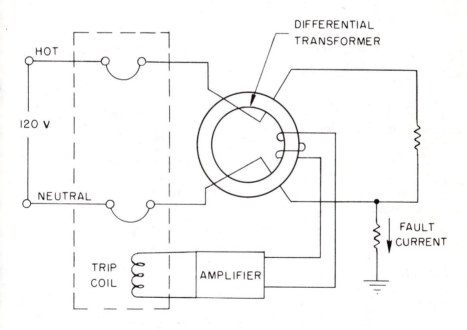

FIG. 4.30. BASIC CIRCUIT OF TYPICAL GFI

EXERCISES

1. 120/240 Volt Service Conductor Loads.

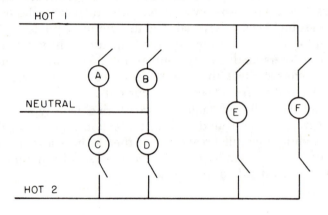

120 V Loads
   A - 12-100 W bulbs
   B - 16 amp electric motor
   C - 15 amp electric heater
   D - 6-200 W heat lamps

240 V Load
   E - 22 amp electric motor
   F - 50 amp electric welder

In the table below, determine the current in each conductor for the particular set of loads connected.

Current in Line (amps)

Loads Connected	Hot 1	Neutral	Hot 2
A			
A & D			
A & C			
A, C & E			
A, B, C, & D			
All			

2. Given a balanced 3ϕ delta system with phase voltages 240 V and phase currents of 25 amps, determine:

(a) the line current and line voltage values,
(b) the total true power output of the system, if the power factor equals 0.90.

3. Given a balanced 3ϕ wye system with phase voltages of 240 V and phase currents of 25 amps, determine:

(a) the line current and line voltage values,
(b) the total true power output of the system, if the power factor equals 0.90.

REFERENCES

ANON., 1974. EHV Transmission Line Routing Manual — Minnesota. Landplan Systems, Commonwealth Associates, Ann Arbor, Mich.

ANON. 1977. National Electric Code 1978. Natl. Fire Prot. Assoc., Boston.

BAYLESTAD, R. L. and L. NASHELSKY. 1977. Electricity, Electronics and Electromagnetics. Prentice-Hall, Englewood Cliffs, N.J.

DOYLE, J. M. 1975. An Introduction to Electrical Wiring. Reston Publishing, Reston, Virg.

KUBALA, T. S. 1974. Electricity 1. Delmar Publishers, Albany, N.Y.

PARADY, W. H. and TURNER, J. H. 1976. Electric Energy. Am. Assoc. for Voc. Instructional Mater., Athens, Ga.

SCHICK, K. H. 1975. Introduction to Electricity. McGraw-Hill Ryerson Ltd., N.Y.

5

Planning the
Farmstead Distribution System

Electrification of farmsteads has enabled farmers to work more efficiently. New equipment and practices, as well as older uses of electricity on the farmstead, continue to produce an ever-increasing demand for electricity. A well-designed farm wiring system will distribute electricity economically and efficiently today as well as be adaptable to future needs.

Characteristics of a well-planned system are that it must be safe, adequate to meet load demands, efficient, and expandable. Compliance with the National Electric Code (NEC) and any existing state or local codes will ensure safety. However, good judgment and proper planning are necessary to meet the other criteria.

An adequate wiring system must have properly-selected components. It must have a main service entrance and building service entrances of sufficient capacities to supply the required power at a high level of efficiency. The building system must have enough branch circuits and outlets of the proper size and type correctly located to meet the farmstead's electrical needs. An efficient system is one which fulfills its needs at the least expense over the lifetime of the system. Three types of costs must be considered: initial cost, maintenance costs, and cost of energy lost in the delivery system. Minimizing initial cost alone may sacrifice operating efficiency in such a way as to increase overall costs. As energy costs increase, minimizing energy loss in the system due to the resistances of the conductors becomes increasingly important.

This chapter will deal with design of electrical distribution systems on a farmstead. The first four sections will deal with selecting and

94

sizing components of a single-phase distribution system. The last sections will deal with three-phase systems for the farmstead.

The first step in designing a single-phase system will be determining the demand load for each building or service area. Using the building loads, location and size of a central distribution center for the system can then be calculated. Finally, the size and type of conductors needed from the central distribution point to the buildings can be established. For a more detailed description of requirements for wiring systems within any particular type of farm building, the reader is referred to such references as Agricultural Wiring Handbook, the Agricultural Engineer's Yearbook, and the National Electric Code.

5.1 DEMAND LOAD FOR FARM BUILDINGS

The first step in designing a farmstead system is to determine the demand load for each building or service area, for loads not within a building. It is not feasible to establish a standard size for each type of building because of the wide variability of farm buildings. Each building must be considered individually. The system suggested has been developed in the NEC (Section 220–40). It is a system for combining known and anticipated loads within a building to determine a demand load by which to size the service for that building. The system first sums all the loads within the building to determine the "total connected load." However, because it is very unlikely that all loads within a building will operate at one time, the NEC suggests a method for determining the "maximum demand load." This demand load is used to size the building service. The demand system developed by the NEC accounts for the diversity of operation within the building.

Making a list of known and anticipated loads is the first step in computing the demand load. Those loads which should be listed are:

(1) Large or permanently connected appliances. The full-load ratings of all equipment of 1500 watts, or ½ horsepower, or greater are generally included in this category. The full-load current of the largest motor should be multiplied by 125% to allow for starting current.

(2) Convenience Outlets. A load of 1.5 amperes at 115 volts should be allotted for each convenience outlet. This accounts for portable tools and appliances.

(3) Lighting Outlets. A load of 1.5 amperes at 115 volts should be allotted for each lighting outlet. Under certain circumstances, such as in poultry laying houses, this figure may be modified to reflect known values either larger or smaller.

The list should be completed by calculating the amperage of 230 volts required to service each load. The amperages converted to this common voltage level will be summed to determine total load. (Recall amperage at 115 volts is converted to amperage at 230 volts by dividing amperes at 115 volts by two).

Note that voltage levels of 115 volts and 230 volts are used to conform with NEC practices. However, 120 volts and 240 volts will be used when referring to service equipment ratings. The latter conforms with equipment suppliers' convention.

Table 5.1 shows the system developed by the NEC (Table 220-40) for determining demand load from the listed loads for buildings with two or more branch circuits. If the load consists of a single item, such as an irrigation pump or crop dryer, the demand system would not be applicable. To apply the system the load without diversity must be determined. The largest combination of loads which are likely to operate at the same time make up the load without diversity. Determining load without diversity requires both a knowledge of the farm operation and good judgment.

TABLE 5.1. FARM BUILDING DEMAND SYSTEM

Demand = Total of		
100%	largest of	1. Load without diversity 2. 125% of largest motor 3. Not less than 60 amp
50%	of	Up to next 60 amp of all other loads
25%	of	Remainder

The following examples will demonstrate the process of listing the loads and applying the demand system.

Example 5.1.

Application of Farm Building Demand System.

Calculate the demand load for a beef barn with a total load of 185 amperes. The largest motor is 3 horsepower (230 volts, single-phase). Load without diversity is 65 amperes.

From Table A.3 of Appendix A—3 hp, single-phase, 230 V motor has a full-load rating of 17 amps.

Applying the demand system,

			Portion of Total	Demand
	1.	Load without diversity = 65 amp	65 amp	65 amp
100% of	2.	Largest motor × 125% =		
largest		17 amp × 1.25 = 21.25 amp		
	3.	60 amp		
50% of		Up to next 60 amp	60	30
25% of		Remainder (185–125)	60	15
			185 amp	110 amp

Demand load = 110 amp at 230 V

Example 5.2.

Demand Load for a Milkhouse.

Determine the demand load for a milkhouse with the following listing of loads.

		Amp at 230 V
Hot water heater	5000 W, 240 V	20.8
Electric furnace	70 amp, 240 V	70.0
Air conditioner	20 amp, 240 V	20.0
Refrigerator	3500 W, 120 V	14.6
Milk transfer pump	1 hp, 120 V	8.0
Bulk milk tank	2½ hp, 240 V	
	(14.5 amp × 1.25)	18.1
19 – 120 V lighting and convenience outlets		14.3
		165.8 amp

The load to operate without diversity for this example was assumed to include the bulk tank, hot water heater, electric furnace, refrigerator, and one-half of the lighting and convenience outlet load for a total of 130.7 amperes.

Applying the demand system,

			Portion of Total	Demand
	1.	Load without diversity = 130.7 amp		130.7 amp
100% of	2.	Largest motor × 125%		
largest		= 18.1 amp		
	3.	60 amp		
50% of		Up to next 60 amp		
		(165.8 - 130.7 = 35.1)	35.1	17.6
			165.8 amp	148.3 amp

Demand load = 148.3 amp at 230 V

When selecting service entrance equipment there are a limited number of standard packages from which to select. They are rated by their amperage capacity at 240 volts. The most common sizes are 30, 60, 100, 150, 200, and 300 amperes. The size selected must be equal to or larger than the calculated demand load. For our milkhouse example, we could select the 150 ampere service entrance. However, if any future expansion is likely, it may be wise to select a larger size to allow for the expansion. It is generally much less expensive to allow extra capacity when first installed than to replace the system later with a larger system.

The smallest service entrance recommended by the NEC is 60 amperes. However, for buildings with only a small amount of load a 30 ampere service as a subservice from another building may be considered.

5.2 CENTRAL METERING AND DISTRIBUTION

The most common type of distribution system on a farmstead has a centrally located distribution center (see Fig. 5.1). Generally the meter will be located at this central or main service location. A service drop or feeder will run from the central point to each building or service area. The central distribution service equipment may consist of up to six circuit breakers or fused switches. One switch may control the wiring serving the distribution center itself and the well pump; the others, up to five, control feeders to the residence and other farm buildings.

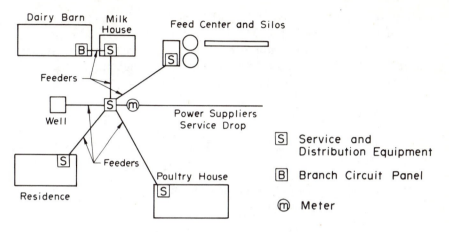

FIG. 5.1.   CENTRAL METERING AND DISTRIBUTION FOR A FARMSTEAD

The central distribution center has a number of advantages:

(1) Safety — Loss of one building will not destroy the system. Service can be maintained to the other buildings. In addition, a separate service drop connected ahead of the building disconnections can be run to the well to ensure a water supply in spite of a fire causing loss of service to any building.

(2) Expandability — When the loads within a building change or a new building is added, feeders to the other buildings are unaffected.

(3) Minimizes main service size — Diversity of load between buildings can be accounted for, thereby minimizing the capacity of the main service needed.

(4) Least investment in wire — Because wires are used very efficiently, this technique minimizes the cost of wire for the system.

(5) Convenience — The meter can be located such that entrance to a building is not required to read the meter and components of the system are easily serviced.

The optimum location for the central distribution point and main service equipment to minimize wiring costs is at the "load center." The load center is the geographic center of the loads. A first step in locating the load center for a farmstead is to draw a scaled map of the farmstead. The location of the service entrance within each building and the demand load for each should be noted. Two perpendicular baselines, designated X and Y, should be located along two sides of the farmstead. The X and Y distance for the load center can now be determined by finding the weighted mean X and Y distances for the loads. The demand loads are used as weighting factors. The distances can be expressed in equation form as

$$Y_{\text{Load Center}} = \frac{\Sigma(L_i Y_i)}{\Sigma L_i} = \frac{\text{Sum of each load} \times \text{its Y-distance}}{\text{Sum of the loads}}$$

$$X_{\text{Load Center}} = \frac{\Sigma(L_i X_i)}{\Sigma L_i} = \frac{\text{Sum of each load} \times \text{its X-distance}}{\text{Sum of the loads}}$$

where L_i = demand load at a building

X_i

Y_i = the coordinates of the building

Σ – implies summation for each building on the farmstead

The following example demonstrates finding the load center.

Example 5.3.

Load Center Calculations.

Locate the load center for the farmstead mapped.

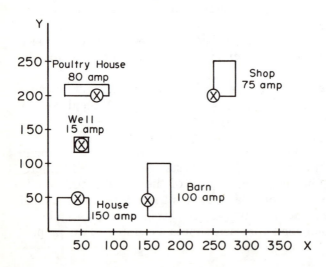

Building	Load	X	X × Load	Y	X × Load
P. House	80	75	6,000	200	16,000
Shop	75	250	18,750	200	15,000
Well	15	50	750	125	2,295
House	150	50	7,500	50	7,500
Barn	100	150	15,000	50	5,000
	420		48,000		45,795

$$X = \frac{\Sigma(L_i X_i)}{\Sigma L_i} = \frac{48,000}{420} = 114 \qquad Y = \frac{\Sigma(L_i Y_i)}{\Sigma L_i} = \frac{45,795}{420} = 109$$

Optimum location for the load center is at X = 114 and Y = 109.

When this method is used to locate the load center, the result must be tempered with other considerations. Topography of the farmstead, location of driveways, trees, buildings, and other obstacles may require the distribution center to be located at a point other than the load center. It may also be necessary to consult the power supplier before locating the distribution center.

5.3 CAPACITY OF MAIN SERVICE

If a centralized distribution system is used, we can take advantage of the diversity between buildings in sizing the main service. It is not likely that all buildings will be operating at their full demand load at the same time. Therefore, Table 5.2 outlines a demand system developed by the NEC (Table 220–41) for calculating the necessary capacity of the main service equipment for the farmstead. The minimum capacity for the main service is the sum of the loads times their appropriate demand factors.

TABLE 5.2. CAPACITY OF MAIN FARMSTEAD SERVICE

Computed Demand Loads	Demand Factor
Residence	100%
All other loads:	
Largest load	100%
2nd largest load	75%
3rd largest load	65%
Sum of remaining loads	50%

Example 5.4.

Sizing Main Service.

Calculate the main service demand load for the farmstead shown in Example 5.3.

Residence	150 amp × 100% = 150 amp
Largest load — Barn	100 amp × 100% = 100
2nd largest load — P. house	80 amp × 75% = 60
3rd largest load — Shop	75 amp × 65% = 49
Remainder — Well	15 amp × 50% = 8
	367 amp

The total minimum demand load = 367 amp at 230 V.

Since service equipment only comes in certain standard rated packages — 100, 200, 300, 400 . . . amp — this farmstead would require a minimum of 400 ampere service.

5.4 SELECTING FEEDER CONDUCTORS

Three major factors which must be considered when selecting conductors for feeder lines or service drops are:

(1) size of wire and insulation type necessary to safely carry the current,

(2) type of wire and insulation needed to meet requirements of its surroundings, and

(3) size of wire necessary to prevent excessive voltage drop in the line.

For the first factor, the NEC has established maximum safe values of current (ampacities) for various wire sizes and insulation types. Operation within the limits of the tables assures the insulation and wire will not be damaged by excessive heat build-up or high temperatures. Tables A.5 and A.6 of Appendix A give the allowable ampacities for various copper and aluminum conductors.

For the second factor, type of wire and insulation needed, we note that the insulation protecting the wire must be suited to the conditions to which it will be exposed. Note that Tables A.5 and A.6 for allowable ampacities are each divided into two sections. The first section of each deals with conductors in free air. The second of each deals with three or less conductors in a raceway, cables, or direct burial of wire. Because of the more favorable heat dissipation conditions, conductors in free air have a higher allowable ampacity than those in cables or buried. For example, a copper No. 10 THW has an allowable ampacity of 40 amperes in free air but only 30 amperes when put underground or in a cable. This means that underground feeders will generally have to be sized larger to supply the same load. Certain insulation types are only suited to particular environments. Table A.7 in Appendix A lists a number of common insulations and their applications. For underground or wet locations the insulated conductors shall be one of the following: RHW, RUW, TW, THW, THWN, lead-covered cable, or aluminum-sheathed cables (ALS). Conductors can be installed in rigid nonmetallic or metallic conduits and raceways of various types. However, cables of one or more conductors of the type mentioned, when directly buried, must have approved outer insulation such as USE or UF.

Note that the size of overhead wire may be limited by the mechanical strength of the wire. According to the NEC, at least No. 10 must be used for spans up to 50 ft and No. 8 for longer spans.

Example 5.5.

Using Allowable Ampacity Tables.

What is the allowable ampacity of:

(a) #10 aluminum, TW insulation, for single conductor in free air?
(b) #10 aluminum, UF insulation, for direct buried?
(c) #8 copper, USE insulation, for direct buried?
(d) #12 copper, THW, cable?

Solution:

(a) 30 ampere (Table A.6)
(b) 25 ampere (Table A.6)
(c) 45 ampere (Table A.5)
(d) 20 ampere (Table A.5)

The third factor, voltage drop in the conductors, is controlled by the resistance of the wires. Resistance of the wires is a function of the cross-sectional area and the length of the wire. As the length of the wire increases, its resistance increases. However, as its cross-section increases its resistance decreases. Tables A.1 and A.2 of Appendix A list the resistance values by wire size for 1000 ft and 1000 m segments of copper and aluminum wires.

Line loss represents the power lost due to the resistance of the wires. It is the product of the current flow through, and voltage drop in, the conductors.

$$P_{Loss} = I E_W$$

where P_{Loss} = line loss in watts

I = current in amperes

E_W = voltage drop in conductors in volts

Line loss is dissipated as heat in the wires and therefore represents a cost which must be controlled by proper wire sizing. The NEC specifies the voltage drop in a branch circuit shall not exceed 3% and that the maximum total voltage drop for feeders and branch circuits shall not exceed 5%. Commonly-used design percentages are 2% for feeders and 3% for branch circuits. Some systems, such as poultry incubators or fluorescent lighting, may require smaller percentage voltage drops.

One method of selecting the correct size of wire is to calculate the allowable resistance for the wire for the design value of voltage drop. The allowable resistance is calculated using Ohm's law in the form,

$$R_A = \frac{E_{Drop}}{I}$$

where R_A = maximum allowable resistance in ohms

E_{Drop} = voltage drop in volts allowable
          = (% voltage drop) $\times E_{Source}$

Example 5.6.

Calculating Allowable Resistance.

Calculate the allowable resistance for a conductor carrying 48 amperes at 240 volts if the allowable voltage drop is 2%.

$$R_A = \frac{0.02 \times 240 \text{ V}}{48 \text{ amp}} = 0.10 \text{ ohms}$$

The total resistance of a conductor is dependent on the length of run and the cross sectional area. Total length of the conductor carrying current will be twice the distance from the source to the load to have a complete circuit.

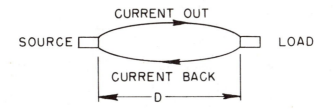

CURRENT OUT

SOURCE   LOAD

CURRENT BACK

D

FIG. 5.2. CONDUCTOR LENGTH FOR A COMPLETE CIRCUIT

The distance from the source to the load is used along with the allowable resistance to calculate the resistance per unit length allowable.

$$\rho = \frac{R_A}{2D}$$

ρ = resistance per unit length in ohms/ unit length

where R_A = allowable resistance in ohms

D = distance from source to load in appropriate length units

Proportions can be used to change the ρ value to a value which can be compared to Tables A.1 and A.2 in Appendix A.

$$R_{1000 \text{ ft}} = \frac{R_A}{2D} \times 1000 \quad \text{or} \quad R_{1000 \text{ m}} = \frac{R_A}{2D} \times 1000$$

A wire size is then selected that has the same or lower R value.

Example 5.7.

Wire Sizing by Allowable Resistance 1.

If the load in Example 5.6 is 80 m from the source, what size copper wire will be needed?

$$R_{1000 \text{ m}} = \frac{0.1 \text{ ohms}}{2 \times 80 \text{ m}} \times 1000 = 0.625 \frac{\text{ohms}}{1000 \text{ m}}$$

From Table A.1 a No. 2 wire with 0.513 ohms/1000 m would be required to maintain less than a 2% voltage drop.

After calculation of wire size by voltage drop, allowable ampacity from Tables A.5 and A.6 should be checked to assure the wire has sufficient ampacity rating.

Example 5.8.

Wire Sizing by Allowable Resistance 2.

Calculate the aluminum wire size needed to maintain a 3% or less voltage drop for a 90 ampere/240 volt load located 220 ft from the central distribution point. Allowable resistance is calculated as:

$$R_A = \frac{E_{Drop}}{I} = \frac{0.03 \times 240 \text{ V}}{90 \text{ amp}}$$

$$= 0.08 \text{ ohms}$$

$$R_{1000 \text{ ft}} = \frac{0.08}{440 \text{ ft}} \times 1000 = 0.1818 \frac{\text{ohms}}{1000 \text{ ft}}$$

From Table A.2 a No. 0 wire with 0.162 ohms/1000 ft would be required. The actual voltage drop using No. 0 wire can be calculated as:

$$R = \frac{0.162 \text{ ohms}}{1000 \text{ ft}} \times 440 \text{ ft} = 0.07128 \text{ ohms}$$

$$E_{Drop} = IR = 90 \text{ amp} \times 0.0712 \text{ ohms}$$

$$= 6.408 \text{ volts}$$

$$\% \text{ Drop} = \frac{6.408}{240} \times 100\% = 2.67\%$$

Combining the information from allowable ampacities, line loss calculations, and insulation type requirements allows the proper selection of conductors. The following two examples will demonstrate the complete process.

Example 5.9.

Wire Selection 1.

What size aluminum wire with TW insulation would be necessary to maintain less than a 2% voltage drop for a 110 ampere 240 volt building service located 125 ft from the load center?

Solution: Calculate allowable resistance per 1000 ft.

$$R_A = \frac{E_{Drop}}{I} = \frac{0.02 \times 240 \text{ V}}{110 \text{ amp}}$$

$$= 0.0436 \text{ ohms}$$

$$R_{1000 \text{ ft}} = \frac{R_A}{2D} \times 1000 = \frac{0.0436 \text{ ohms}}{2 \times 125 \text{ ft}} \times 1000$$

$$= 0.174 \text{ ohms}$$

From Table A.1, use No. 0 (with $0.162 \dfrac{\text{ohms}}{1000 \text{ ft}}$).

From Table A.6, allowable ampacity for No. 0 with TW insulation is 150 amperes. Therefore, No. 0 with TW is adequate.

Example 5.10.

Wire Selection 2.

What size copper wire will be needed in a cable (UF insulation) to be placed underground if it is to supply a 90 ampere/240 volt load 30 m from the source? Assume a 2% voltage drop.

Calculate the allowable resistance per 1000 m.

$$R_A = \frac{E_{Drop}}{I} = \frac{0.02 \times 240 \text{ V}}{90 \text{ amp}}$$

$$= 0.0533 \text{ ohms}$$

$$R_{1000 \text{ m}} = \frac{R_A}{2D} \times 1000 = \frac{0.0533 \text{ ohms} \times 1000}{2 \times 30 \text{ m}}$$

$$= 0.888 \frac{\text{ohms}}{1000 \text{ m}}$$

From Table A.1 — to maintain the allowable voltage drop a No. 4 wire would be required. Checking Table A.5 for allowable ampacity, a copper cable, underground, would require a No. 2 UF to maintain allowable ampacity. Therefore, a No. 2 UF would be required.

The methods outlined above apply to circuits and feeders if 120 volt or 240 volt, two-wire or three-wire. For the case of three-wire feeders, on 120/240 volt feeders, the size of the neutral conductor may be figured on the maximum unbalanced load between either ungrounded conductor and the neutral, omitting all 240 volt loads, unless prohibited by local regulations.

5.5 THREE-PHASE FARMSTEAD SERVICES

The decision as to whether three-phase power should be used for all, part or none of an electrical installation depends on electrical and economic considerations. Generally on farmsteads the larger individual loads tend to be motor loads. The initial cost of three-phase motors to power these loads is considerably less than single-phase motors. It would appear logical to use three-phase motors whenever possible since they are less expensive to purchase. However, the situation is not that simple. It usually costs the power supplier more to furnish three-phase service since additional transformers and lines are necessary. Therefore, the rate for three-phase power may be higher than single phase. Operating costs may destroy the benefit of lower initial costs for the motor. In many regions most of the power distribution lines in the rural area are single-phase systems. The power company may be unwilling to rebuild the distribution to the farmstead to make three-phase power available. Even though three-phase power may not be available, the use of three-phase equipment may still be possible with the addition of phase converters. A phase converter is a device which will allow a three-phase motor to operate from a single-phase source. Phase converters are discussed in Section 5.6.

Consideration of using three-phase power should be made when significant new large motor loads are being added, such as when adding a grain drying system or an irrigation system, or perhaps when a significant amount of rewiring or replacing of motors is to be done. It is advisable in all cases to consult the power supplier in order to obtain all pertinent information and to obtain bids from an electrical contractor. With these inputs a sound decision can be made.

The fundamentals of designing a three-phase system are no different than those developed for a single-phase system. Location of the service and sizing of components follows the same type of procedure. Three common types of three-phase and combined three-phase, single-phase systems are described in the following sections. In some installations, such as irrigation equipment, all equipment may be powered by three-phase system; therefore, no need would exist for a single-phase system. However, on most farmsteads the three-phase system would supply large motor loads and a second single-phase system would supply lighting and all the other smaller loads.

5.5.1 120/240 Volt Four-Wire, Three-Phase, Delta

This type of service is shown schematically in Fig. 5.2. Three transformers are used in a delta configuration. One of the transformers is generally sized larger than the others because it supplies not only one-third of the three-phase power, but also the single-phase power. This transformer has a center tap connected to the grounded neutral of the single-phase system. This system has the advantage of requiring only one service drop. However, it has the disadvantage of unbalancing the three-phase system if the single-phase loads are large in proportion to the three-phase loads.

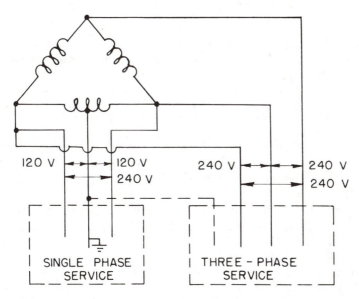

FIG. 5.2. 120/240 VOLT, FOUR-WIRE, THREE-PHASE DELTA SYSTEM

5.5.2 Separate 120/240 Volt Single-Phase and Three-Wire, Three-Phase System

This system is actually two isolated distribution systems. Four transformers are required. One transformer supplies the 120/240 volt single-phase service. The other three are used to make up the wye or delta three-phase system.

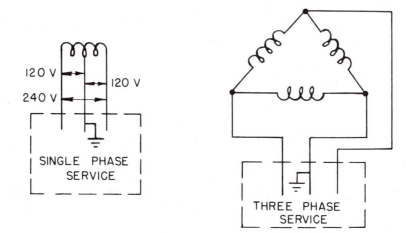

FIG. 5.3.   120/240 V SINGLE-PHASE AND THREE-PHASE DELTA

This system requires two separate service drops. However, it allows the three-phase system to be at a higher voltage level, such as 480 volt, for large equipment.

5.5.3 120/240 Volt, Four-Wire, Three-Phase Open Delta

This system is operated with only two transformers. Service is provided from a V-phase primary line (two-phase wires). One of the transformers is larger since it supplies the single-phase and a portion of the three-phase.

This service, like the four-wire, closed-delta, has the advantages of requiring only one service drop and yielding standard voltage levels for both single and three-phase loads. It also has the disadvantage of unbalancing the three-phase system if the single-phase is large in comparison to the three-phase loads. In addition, three-phase motors cannot be as heavily loaded with this system without unbalancing the motor winding currents.

For this system the combined capacity of the two transformers

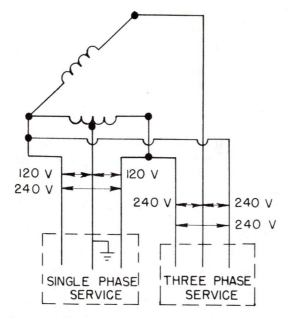

FIG. 5.4. 120/240 VOLT, FOUR-WIRE, THREE-PHASE OPEN-DELTA SYSTEM

must be at least 115% of the combined capacity of the three trans-formers in a closed delta system to supply the same load.

5.6 PHASE CONVERTERS

Phase converters make it possible to operate three-phase motors from single-phase power lines. They convert single-phase line voltage into a three-phase system. Converter-motor combinations are widely used for many kinds of farm loads. Some examples are crop driers, grain handling systems, irrigation pumps and animal feeding systems.

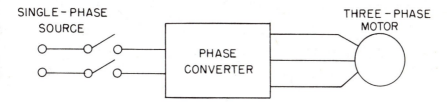

FIG. 5.5. PHASE CONVERTER-MOTOR COMBINATION

Before an investment is made in a phase-converter, careful analysis should be made to make sure a converter-motor combination is the best approach.

Some situations where a phase-converter may be the best choice include:

(1)   when three-phase power is needed, but the cost of bringing the three-phase to the farmstead is prohibitive because of power line construction costs.

(2)   when cost per kilowatt hour is considerably larger for three-phase than single-phase power. In this case the savings in operating might quickly pay for the converter.

(3)   when you need to use a large motor, but the starting current of the single-phase motor is too high for the line to handle. Power suppliers often limit the horsepower of a motor which is supplied by a single-phase line. The high starting current of a large single-phase motor may reduce the line voltage and adversely affect the service to other customers and equipment. However, converter-motor combinations generally draw less starting current for the same size of motor. Consequently, power suppliers may allow the use of higher horsepower through a converter.

(4)   when temporary three-phase power is needed until regular three-phase service becomes available.

(5)   when a number of motors are needed and the cost of three-phase motors and a converter is smaller than the cost of single-phase motors.

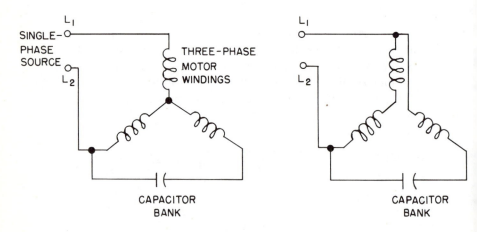

FIG. 5.6. SIMPLIFIED DIAGRAMS OF THE TWO DESIGNS FOR CAPACITOR CONVERTERS

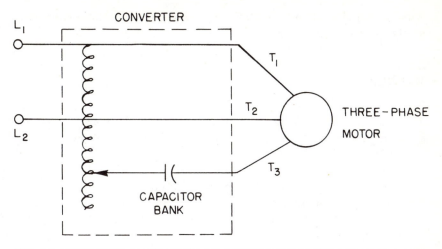

FIG. 5.7. SIMPLIFIED DIAGRAM OF AN AUTOTRANSFORMER CAPACITOR CON-
VERTER

Two general types of converters are available — static and rotary. Each has advantages under specific conditions. Static converters are available in capacitor and autotransformer capacitor designs. The static converters, shown diagramatically in Fig. 5.6 and 5.7, have no moving parts, hence the term "static".

A rotary converter consists of a unit which appears similar to a motor and a bank of capacitors. Figure 5.8 shows a simplified diagram of a rotary converter.

Further details about selection and installation of phase converters should be sought before a decision is made. The last refer-

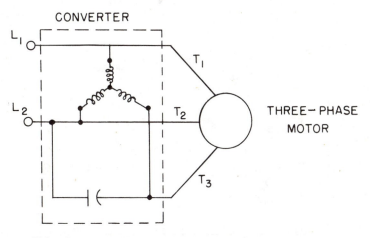

FIG. 5.8.   SIMPLIFIED DIAGRAM OF A ROTARY CONVERTER

ence at the end of this chapter is a good starting point for continued study of phase converters.

EXERCISES

1. Determine the demand load for a building, which is used for machinery storage and as a workshop with the following listing of loads:

   18–120 V lighting and convenience outlets
   3.5 kW electric heater
   1 hp automatic air compressor
   60 amp, 240 V welder

2. Determine the demand load for a milkhouse with the following loads:

   10–120 V lighting and convenience outlets
   bulk tank 3 hp compressor
          1/3 hp agitator
   milk transfer pump – 3/4 hp
   water heater 3 kW
   refrigerator 1.5 kW
   ventilation fan 1/3 hp

3. Calculate the load center for the farmstead shown below.

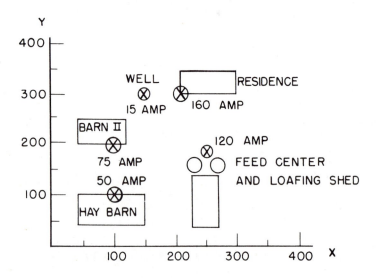

4. Calculate the main service demand load for the farmstead in problem 3.
5. What size service entrance would you recommend for each building in problem 3?
6. What size and type of copper conductors would you recommend for each of the necessary service drops for the farmstead in problem 3 if a) underground feeders are used, b) overhead feeders are used.

REFERENCES

ANON. 1976. Agricultural Wiring Handbook. Farm Electrification Council, Columbia, Mo.

ANON. 1977. National Electric Code 1978. Natl. Fire Prot. Assoc., Boston.

ASAE. 1978. Agricultural Engineers Yearbook. Am. Soc. of Agric. Eng., St. Joseph, Mo.

BUTCHBAKER, A. F. 1977. Electricity and Electronics for Agriculture. Iowa St. Univ. Press, Ames, Ia.

RICHEY, C. B., JACOBSON, P., and HALL, C. W. 1961. Agricultural Engineers' Handbook. McGraw-Hill, N.Y.

SODERHOLM, L. H. 1972. Phase converters for operation of three-phase motors from single-phase power. USDA, Farmers Bulletin No. 2252, Washington, D.C.

6

Residential Electrical Planning

This chapter is intended as a guide for planning an electrical system for a new home or evaluating the needs of an existing home. In an existing home, a detailed inspection of the present electrical system will be needed. In a new home, plans and specifications can be used. In this chapter we will deal primarily with number and location of outlets needed, type and number of branch circuits needed and sizing the service entrance equipment.

Electrical power reaches a home through service conductors (the wires that extend from the utility distribution system to the house, pass through the meter and end at the service equipment). The service equipment usually consists of a main entrance switch, fuses or circuit breakers, followed by a distribution panel with overcurrent protective devices for each branch circuit. Branch circuits are those circuits connecting the final overcurrent protection with the loads. Controls such as switches are another essential part of the wiring system which are contained with the branch circuits. A grounding conductor connects the service neutral to a grounding electrode, such as a cold water pipe or driven ground rod. Figure 6.1 summarizes the delivery system.

In most locations the electrical system for a residence must meet the National Electric Code (NEC) requirements or the requirements of any existing state or local codes which may supersede the NEC. The purpose of these codes is to safeguard people and property from hazards arising from the use of electricity. These codes establish a minimum standard acceptable for safety. A system meeting code requirements should be essentially hazard-free but will not necessarily be efficient, convenient, or adequate for good service or future

116

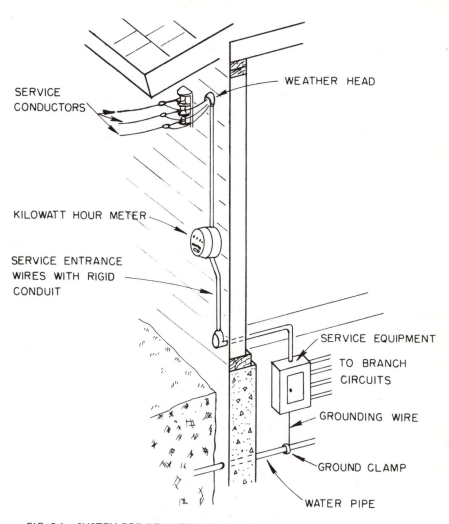

SERVICE CONDUCTORS

WEATHER HEAD

KILOWATT HOUR METER

SERVICE ENTRANCE WIRES WITH RIGID CONDUIT

SERVICE EQUIPMENT

TO BRANCH CIRCUITS

GROUNDING WIRE

GROUND CLAMP

WATER PIPE

FIG. 6.1. SYSTEM FOR DELIVERY OF ELECTRICAL POWER TO A RESIDENCE

expansion. Therefore, good judgment and other guidelines will be necessary in designing efficient and convenient systems.

6.1 ELECTRICAL SYMBOLS FOR PLANS AND BLUEPRINTS

Electrical symbols that commonly appear on the architectural drawings of a residence are shown in Fig. 6.2. It is advantageous

Lighting Outlets

○   Ceiling Light Fixture

–○   Wall Mounted Light Fixture

Ⓡ   Recessed Light Fixture

Ⓢ   Ceiling Pull Switch
     Light Fixture

Fluorescent Light Fixture

Continuous Row of Fluorescent
Fixtures

Cluster Swivel Lamp Holder

Receptacles and Special-Purpose Outlets

⊖   Duplex Receptacle Outlet

⊖   Duplex Receptacle Outlet,
     Split Wired

⊖ R   Range Outlet

Ⓕ   Fan Outlet

Ⓙ   Junction Box

–Ⓒ   Clock Outlet

$-\triangle^X$ Single Special-Purpose
       Receptacle Outlet (Superscript
       X for reference to description)

$\triangle^X$ Duplex Special-Purpose
       Receptacle Outlet (Superscript
       X for reference to description)

–▲   Special-Purpose Outlet or
      Connection (DW – Dish Washer,
      CD – Clothes Dryer, GD –
      Garbage Disposal, etc.)

Switches

S   Single-pole Switch

S_3 Three-way Switch

S_4 Four-way Switch

S_K Key-operated Switch

S_P Switch and Pilot Light

S_D Door Switch

S_T Timer Switch

Identification Subscripts for Outlets, Fixtures, and Switches

WP   Water Proof

VT   Vapor Tight

WT   Water Tight

RT   Rain Tight

G   Grounded

DT   Dust Tight

EP   Explosion Proof

R   Recessed

GF   Ground Fault Protected

Miscellaneous

▨▨▨   Service Panel

████   Distribution Panel

– – – –   Switch Control Indicator.
          Connects Outlets or Lights with
          Control Points.

FIG. 6.2.   ELECTRICAL SYMBOLS FOR PLANS AND BLUEPRINTS

to be familiar with these symbols and to make use of standard symbols when doing planning work.

6.2 PLACEMENT OF OUTLETS AND SWITCHES

For maximum convenience as well as safety it is necessary to have an adequate number of outlets and switches. These outlets must be carefully located and of the correct type to meet present and future needs.

After studying the following sections which describe the general requirements for outlets in a residence, the "Residential Electric Outlet Planning Guide" included as Appendix B may be helpful in planning outlets for specific rooms.

6.2.1 General Receptacle and Appliance Outlets

Living rooms, bedrooms and other general living areas require convenient outlets for lamps, radios, television and other portable appliances. Enough receptacles should be provided to avoid use of extension cords, particularly across travel paths. The NEC specifies that in these general living areas no point along the floor line in any wall space should be more than 1.8 m (6 ft) from a receptacle outlet and that any wall space greater than 0.6 m (2 ft) should have an outlet. Duplex-receptacle 120 volt outlets, as shown in Fig. 6.3, are most commonly used to meet these requirements.

Greater accessibility is provided when the receptacles are located with possible furniture placement in mind. For example, considering

FIG. 6.3. DUPLEX RECEPTACLE, 120 V

a 3.6 m (12 ft) wall section in a living room, one receptacle centrally located would meet the minimum requirements. However, it would likely be blocked by a heavy piece of furniture. Locating receptacles near the ends of the space may be a better plan.

Although not required by the code, at least one convenience outlet should be placed in each hall area for cleaning equipment and night lights. The only convenience outlet required in the bathroom is one located convenient to the mirror area. To avoid unsafe situations, no outlets should be placed in or near tubs or showers.

Although NEC only requires one outlet for an unfinished basement area, more would be desirable. One receptacle in the laundry area should be located near water and drain connections for a washing machine. Another receptacle should be available for ironing and other laundry room activities.

It is recommended that appliance-circuit receptacles be spaced every 0.3 m (1 ft) to 1 m (3 ft) along kitchen or dining room counter areas. NEC requires an outlet for a counter space longer than 0.3 m (1 ft).

6.2.2 Lighting Outlets

The NEC requires that at least one wall switch controlled lighting outlet shall be installed in every habitable room, in hallways, stairways, and attached garages, and at outdoor entrances. At least one lighting outlet (not necessarily wall switch controlled) is required in each attic, underfloor space, utility room and basement. One exception is permitted, that is in habitable rooms other than kitchens or bathrooms, one or more switched receptacle is acceptable in lieu of the lighting outlet.

When considering location of lighting controls it is important to remember when walking through the house, it should always be possible to light the path ahead. Similarly, it should always be possible to turn off the lights without retracing steps. This also applies to outdoor routes between house and garage or other buildings.

The type and number of lighting outlets required to maintain recommended levels of illumination in interior areas is discussed in detail in Sections 9.3 through 9.5 of this text.

6.2.3 Special Purpose Outlets

Special purpose outlets, often at 240 volts, are required for large or special appliances. Appliances such as ranges, air conditioners, water heaters, furnaces, and dishwashers will require special purpose outlets. Appropriate symbols from Fig. 6.2 should be used to identify these outlets and distinguish them from the general receptacle outlets.

6.3 BRANCH CIRCUITS

The next step in planning a residential electrical system is establishing the number and size of branch circuits needed. Branch circuits, as defined earlier, originate either in the service entrance equipment or from subdistribution panels which are, in turn, supplied by feeders from the service equipment.

Normally, each 120 volt circuit will have one fuse or circuit breaker located in series with the hot conductor in the service panel. A 240 volt circuit requires two fuses or two circuit breakers linked together.

In order to operate efficiently, branch circuits must be of sufficient size to handle the connected load. When a branch circuit becomes overloaded, excessive power loss occurs. This power loss is dissipated as heat in the wires and may create a fire hazard. Common signs of overloaded branch circuits include 1) fuses "blowing" or circuit breakers needing resetting frequently, 2) heating appliances never reaching desired temperature, 3) motors overheating, and running slowly, or 4) television picture shrinking when appliances are in use.

Branch circuits are classified by the amperage rating of their overcurrent protection. Common sizes of 15, 20, 30, 40 and 50 amperes are used in residential wiring. The size of wire used must have sufficient ampacity rating to carry the full current (see Tables A.5 and A.6 of Appendix A). For example, minimum copper wire sizes for 15 and 20 ampere branch circuits are #14 and #12 respectively.

However, a branch circuit using #12 copper wire with a 15 ampere overcurrent protection device is still classified as a 15 ampere branch circuit.

Branch circuit needs are most easily determined by looking at needs in three categories: 1) lighting and general-purpose circuits, 2) small appliance circuits, and 3) special purpose circuits.

6.3.1 Small Appliance Circuits

Small appliance circuits provide power to receptacles for portable appliances in the kitchen, dining area and laundry. No lighting may be connected to these circuits. The NEC requires a minimum of two 20 ampere, 120 volt, small appliance circuits to serve kitchen and dining areas. A third small appliance circuit is required to the laundry. Only 20 ampere circuits may be used as small appliance circuits.

6.3.2 Lighting and General-Purpose Circuits

Lighting and general-purpose circuits supply lighting outlets throughout the house and convenience outlets except in the kitchen, dining area, and laundry. These circuits can be either 15 ampere or 20 ampere, 120 volt circuits. The NEC requires a minimum of one 20 ampere circuit for each 46 m^2 (300 ft^2) or one 15 ampere circuit for each 35 m^2 (375 ft^2). The NEC minimum may require more outlets per circuit than is reasonable for adequate service. A better recommendation is to allow no more than 8 to 10 outlets for each 15 ampere circuit and no more than 10 to 12 for each 20 ampere circuit.

Example 6.1.

Branch Circuits for General Lighting and Convenience Outlets.

A residence is determined to need 70 lighting and convenience outlets, not including those to be supplied by the small appliance circuits. What type and how many branch circuits are needed to serve this load? Several alternatives exist. Three alternatives are:
Alternative 1 (all 15 amp circuits):

$$\frac{70 \text{ outlets}}{\dfrac{10 \text{ outlets}}{\text{circuit}}} = 7 \text{ circuits minimum}$$

Alternative 2 (all 20 amp circuits):

$$\frac{70 \text{ outlets}}{\dfrac{12 \text{ outlets}}{\text{circuit}}} = 5.8, \text{ therefore need 6 circuits minimum}$$

Alternative 3 (combination of 15 and 20 amp circuits):

Five 20 amp circuits	60 outlets
One 15 amp circuit	10 outlets
	70 outlets

Outlets supplied by these circuits should be divided as evenly as possible between circuits. It is recommended that each room be supplied more than one circuit to diversify load. The NEC code requires that all receptacles in the bathroom, laundry, and those outdoors must be on circuits which have ground-fault interrupters.

6.3.3 Special Purpose Circuits

Special purpose circuits serve individual permanently installed appliances or loads. The usual procedure is to provide a branch circuit for each of the following loads:

(1) Electric range
(2) Automatic clothes washer
(3) Clothes dryer
(4) Electric water heater
(5) Dishwasher
(6) Garbage disposal
(7) Water pump
(8) Motor on oil-burning furnace
(9) Motor on blower in furnace
(10) Permanently connected motors with ratings greater than 1/8 horsepower
(11) Permanently connected appliances with ratings greater than 1000 watts

The circuits can be either 120 or 240 volts depending on the load requirements. For the larger loads 240 volts are preferred because smaller wires can be used. Nameplate ratings of appliances should be used whenever possible. However, if nameplate ratings are not available, Table 6.1 may be used to estimate the appliance size.

TABLE 6.1. ESTIMATED APPLIANCE LOADS

Average Wattages for Household Equipment	
Air Conditioner – Room	1,400 watts
Automatic Washer	700
Attic Fan	500
Clothes Dryer – Gas	350
– Electric	5,000
Compactor	1,200
Dishwasher	1,200
Disposal Unit	500
Furnace – Gas	100
– Oil	800
Heater – Wall	1,250
– Bathroom	1,500
Range	12,000

Some of these branch circuits may also serve other loads or receptacles. However, when a branch circuit serves more than one load or receptacle, the circuit capacity is limited. When the circuit supplies both portable and permanently connected appliances, the permanently connected load must not exceed 50% of the circuit rating. In addition, any one portable appliance on a circuit should not exceed 80% of the rated circuit capacity. Therefore, no appliance over 12 amperes or 1,380 watts should be supplied by a 15 ampere circuit, no appliance over 16 amperes or 1,840 watts on a 20 ampere circuit.

Branch circuits which are rated at 30, 40, or 50 amperes cannot be used for lighting or small appliance circuits in residences. Their use is limited to fixed appliances such as electric water heaters and fixed space heaters. The NEC specifies that the maximum load allowable on a 30 ampere circuit should not exceed 80% of its rating, 24 amperes.

Sizing the branch circuit for service to an electric range presents a special situation. Because it is unlikely that all portions of the range will be in use at their full capacity at the same time, sizing the service for the maximum load the range could produce would require a larger branch circuit and wires than is really necessary. The NEC has developed a demand factor system for electric ranges. The demand load is that portion of the maximum load or nameplate rating for which the range branch circuit should be sized.

The NEC system requires that for ranges up to 12 kilowatts a demand load of 8 kilowatts should be used. For ranges between 12 kilowatts and 27 kilowatts, 0.2 kilowatts should be added to the demand load for each one kilowatt or major fraction thereof over 12 kilowatts. For homes with more than one range or ranges larger than 27 kilowatts, Table 220-19 of the NEC should be consulted.

Example 6.2.

Branch Circuits for Electric Ranges.

What size branch circuits are needed for ranges with the following rated maximum loads?

(a) 16.2 kW
(b) 11 kW
(c) 20.6 kW

(a) Maximum load = 16 kW
Demand Load = 8 kW + (16 - 12) (0.2 kW)
= 8.8 kW

$$\frac{8,800 \text{ W}}{240 \text{ V}} = 36.6 \text{ amp} \qquad \text{Therefore, need 40 amp circuit}$$

(b) Maximum load = 11 kW
Demand load = 8 kW

$$\frac{8,000 \text{ W}}{240 \text{ V}} = 33.3 \text{ amp} \qquad \text{Therefore, need 40 amp circuit}$$

(c) Maximum load = 20.6 kW
Demand load = 8 kW + (21-12) (0.2 kW)
= 9.8 kW

$$\frac{9,800 \text{ W}}{240 \text{ V}} = 40.8 \text{ amp} \qquad \text{Therefore, need 50 amp circuit.}$$

As can be seen from the discussion above, there is no standard number of branch circuits required for a residence. The number needed will vary with the size of the home and the number of electrical appliances. As a general rule, it is wise to use more circuits than is required by the NEC. The presence of extra circuits permits the addition of other electrical equipment in the future.

Example 6.3.

Determining Branch Circuits Requirements.

Determine the branch circuit requirements for the home diagramed below. Assume the following major appliances with given nameplate ratings:

Garbage disposal - 120 V	600 W
Dishwasher - 120 V	1,300 W
Range - 120/240 V	14,000 W
Water heater - 240 V	3,000 W
Clothes dryer - 240 V	5,000 W
Furnace fan - 120 V	500 W
Bath fan (1/8 hp) - 120 V	400 W
Air conditioner - 240 V	4,500 W

MAIN FLOOR PLAN

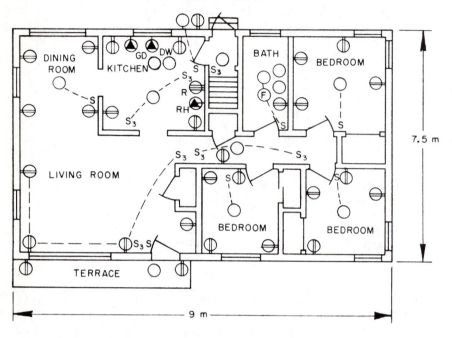

BASEMENT PLAN

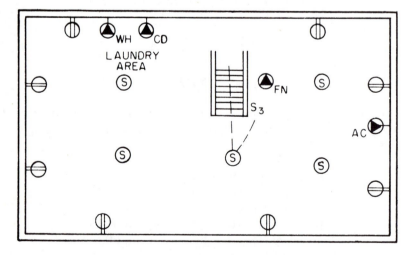

Number of outlets to be served by general lighting and convenience circuits = 47.

Listing of Branch Circuits

Circuit. No.	Load Served Load Served	Circuit Rating
1-6	General lighting and convenience outlets	15 amp/120 V
7,8	Small appliance – kitchen and dining	20 amp/120 V
9	Small appliance – laundry	20 amp/120 V
10	Garbage disposal	15 amp/120 V
11	Dishwasher	15 amp/120 V
12	Range (demand = 8.4 kW)	40 amp/240 V
13	Water heater	20 amp/240 V
14	Clothes dryer	30 amp/240 V
15	Furnace fan	15 amp/120 V
16	Bath fan	15 amp/120 V
17	Air conditioner	30 amp/240 V
18-20	Spares for expansion	

6.4 SIZING THE SERVICE ENTRANCE

Once the loads within the residence have been established, the size of the service entrance equipment needed can be determined. Since it is highly unlikely that all loads will be in use at one time, the NEC presents two methods for determining demand load for the residence. The more general method of the two (NEC 220-10 B.) will be presented here.

This approach divides the load into four categories:

(1)   General lighting and small appliance
(2)   Electric range
(3)   Heating and air conditioning
(4)   Other appliances.

The total demand load will be the total of the load from each of these four categories.

6.4.1 General Lighting and Small Appliances

Total general lighting and small appliance load will represent the sum of load for lighting, convenience receptacles, and small appliance circuits. General lighting load is determined by the area of the residence. A factor of 32 watts/m^2 (3 watts/ft^2) is used for all floor area excluding open porches, garages, or unused or unfinished spaces not adaptable for future use. A load of 1,500 watts is assumed for each small appliance circuit in the residence.

Example 6.4.

Total General Lighting and Small Appliance Load.

What is the total general lighting and small appliance load for a two-story residence, with full basement? Outside dimensions of the home are 7.3 m (24 ft) by 9.1 m (30 ft). Assume three small appliance circuits.

General Lighting Load = 7.3 m × 9.1 m

$$\times \ 3 \ \text{floors} \times \frac{32 \ \text{W}}{\text{m}^2} \qquad\qquad = \ 6{,}377 \ \text{W}$$

three small appliance circuits × 1,500 W/circuit   = 4,500 W

Total   = 10,877 W

The NEC also presents a demand system to calculate the demand load for this category from the total load. The demand load is calculated by applying the demand system in Table 6.2.

TABLE 6.2.   GENERAL LIGHT AND SMALL APPLIANCE DEMAND
SYSTEM FOR RESIDENCE

Portion of Load to Which Demand Factor Applies (Watts)	Demand Factor (%)
First 3,000 or less at	100
Next 3,001 to 120,000 at	35
Remainder over 120,000 at	25

Example 6.5.

Demand for General Light and Small Appliance.

Calculate the demand load for general lighting and small appliances for the total load calculated in Example 6.4.

$$3,000 \text{ W} \times 100\% = 3,000 \text{ W}$$
$$(10,877 - 3,000) \text{ W} \times 35\% = \underline{2,757 \text{ W}}$$

Total Demand = 5,757 W
for General
Light

6.4.2 Electric Range

If an electric range is to be installed in the residence, the demand load used in calculating the service entrance needs for the range is the same as the procedure used for sizing the branch circuit serving the range (see Section 6.3.3). Ranges up to 12 kilowatts have a demand load of eight kilowatts. For each one kilowatt over 12 kilowatts, 0.2 kilowatts is added to the eight kilowatts to get the range demand load.

6.4.3 Heating and Air Conditioning

Heating and air conditioning are included in the same category because of the diversity between the two types of loads. That is, the two types of loads will not be operating at the same time; therefore, only the larger of the two will be used in calculating service entrance capacity needed. In this category heating would include electric space heating and fuel fired furnaces; air conditioning includes both central and room air conditioners.

Example 6.6.

Determining Heating and Air Conditioning Load.

Calculate the demand load to be included in service entrance calculations for a home with 2.5 kilowatts electric space heat and two 10 ampere/240 volt room air conditioners.

Heating Load   2,500 W

A. C. Load   2 × 10 amp × 240 V = 4,800 W

The larger load 4,800 W would be used in calculation of service entrance requirements.

6.4.4 Other Appliances

Any appliance large enough to justify a special purpose branch circuit, as described in Section 6.3.3, and not already accounted for in the other three load sections should be included in this section. If four or more fixed appliances are included, a demand factor of 75% can be used on the total of the fixed appliances. Fixed appliances are defined as those appliances which are fastened or otherwise secured in a specific location, such as a water heater or a garbage disposal. The NEC also specifies a minimum of 5000 watts be included if an electric clothes dryer is to be installed.

The total load from this section is then the total of the appliance loads with the 75% demand factor on fixed loads if four or more are included.

Example 6.7.

Other Appliance Loads.

Calculate the load to be included in service entrance load for a home with the following listing of appliances.

Range	17,200 W	Dishwasher (built-in)	1,500 W
Furnace	700	Water heater	4,500
Electric heat	2,000	Sump pump	1,000
Central air conditioning	10,000	Garbage disposal	1,500
		Clothes dryer	5,200

Listing those appliances which would not be covered in one of the other three categories,

Fixed		*Non-Fixed*	
Dishwasher	1,500 W	Clothes dryer	5,200 W
Water heater	4,500		
Garbage disposal	1,500		
Sump pump	1,000		

8,500 × 0.75 = 6,375 W

Total load from category of other appliances = 6,375 W + 5,200 W = 11,575 W.

6.4.5 Standard Service Entrance Sizes

The service entrance selected must have an ampacity rate equal to or greater than that calculated from the procedures outlined. Service entrance equipment packages are available only in a limited number of amperage ratings. The common service entrance sizes are 60, 100, 150 and 200 amperes.

A 60 ampere service is the minimum acceptable according to the NEC. This generally provides sufficient capacity for lighting and portable appliances, including only one major electrical appliance such as range, clothes dryer or water heater. This service is not recommended for houses over 93 m^2(1,000 ft^2). A 100 ampere service is the required minimum for new homes in many areas. A 150 ampere service is desirable in many modern homes with larger electrical appliances. A 200 ampere service or even larger may be required when electric resistance heating is to be used in the home.

Example 6.8.

Selecting a Standard Service Entrance Size.

Select a service entrance size, provided the total demand load for the service entrance is:

(a) 22,000 W
(b) 30,000 W
(c) 13,000 W

(a) $\dfrac{22,000 \text{ W}}{240 \text{ V}} = 91.8 \text{ amp}$ Need 100 amp service.

(b) $\dfrac{30,000 \text{ W}}{240 \text{ V}} = 125 \text{ amp}$ Need 150 amp service.

(c) $\dfrac{13,000 \text{ W}}{240 \text{ V}} = 54.2 \text{ amp}$ Need 60 amp service.

Example 6.9.

Calculating Service Entrance Size.

Calculate the service entrance size needed for the residence described in Example 6.3.

(1) General lighting and small appliance load

$$\text{General light load} = 2 \times 7.5 \text{ m} \times 9.0 \text{ m} \times \frac{32 \text{ W}}{\text{m}^2} = 4320 \text{ W}$$

$$\text{Three small appliance circuits} \times \frac{1500 \text{ W}}{\text{Circuit}} \qquad = 4500 \text{ W}$$

$$\overline{\hphantom{xx}8820 \text{ W}}$$

$$\text{Applying demand system } 3000 \text{ W} \times 100\% \qquad = 3000 \text{ W}$$
$$5820 \text{ W} \times \quad 35\% \qquad = 2037 \text{ W}$$

$$\overline{\hphantom{xx}5037 \text{ W}}$$

(2) Electric range
14,000 W Electric range
Demand load (Section 6.3.3) = 8400 W

(3) Heating and air conditioning
Heating – furnace   500 W
Air conditioning   4500 W
Larger is air conditioning at 4500 W

(4) Other appliances

Fixed		*Non-Fixed*	
Garbage disposal	600 W	Clothes dryer	5000 W
Dishwasher	1300		
Water heater	3000		
Bath fan	400		

$$5300 \text{ W} \times 0.75 = 3975 \text{ W} + 5000 \text{ W}$$
$$= 8975 \text{ W}$$

Demand load for service entrance
General light and small appliances   5037 W
Electric range   8400
Heating and air conditioning   4500
Other appliances   8975

$$\overline{\hphantom{xxx}26{,}912 \text{ W}}$$

$$\frac{26{,}912 \text{ W}}{240 \text{ V}} = 112 \text{ amp} \qquad$$ Therefore, need 150 amp service entrance.

EXERCISES

1. Describe three possible combinations of circuits to supply a general lighting and convenience load with:

   (a) 75 outlets
   (b) 60 outlets

2. What size branch circuit is required for each of the following electric range sizes:

   (a) 7.5 kW
   (b) 14.4 kW
   (c) 18.6 kW
   (d) 16 kW

3. Make a listing of branch circuits required for a home with (2833 ft^2) of floor area. The house has a minimum number of small appliance circuits.

Appliances Within the Home	Nameplate Rating
Range	17,800 W
Water heater	4,000
Clothes dryer	3,000
Furnace	400
Air conditioner (central)	4,800

4. Calculate the general lighting load for each of the following:

   (a) Two-story house 25 ft by 40 ft, no garage, no basement.
   (b) A single-story house 10 m by 15 m, with a 7 m by 7.5 m attached garage. The house has a full basement, half of which was finished for a rec-room when constructed. The other half is unfinished but may be finished in the future.

5. For a dwelling which has a floor area of 140 m^2 exclusive of unoccupied cellar, unfinished attic and open porches, calculate the service entrance needed. It has a 12 kW range, a 2.5 kW water

heater, a 1.2 kW dishwasher, a 9 kW electric space heater, a 5 kW clothes dryer, and a 6 amp, 240 V room air conditioner unit for appliances.

REFERENCES

ANON. 1977. National Electric Code 1978. Natl. Fire Prot. Assoc., Boston.
DOYLE, J. M. 1975. An Introduction to Electrical Wiring. Reston Publishing, Reston, Va.
HERRICK, C. N. 1975. Electrical Wiring: Principles and Practices. Prentice-Hall, Englewood Cliffs, N.J.
JONES, R. A. and SPIES, H. R. 1964. Electric Wiring. Small Homes Council. Circular Series G4.2. Univ. of Ill. Urbana, Ill.

7

Electrical Controls

Control devices allow one to regulate the flow of electrical energy and thereby control electrical equipment such as lights, motors, generators and heaters. Electrical controls have made a tremendous contribution towards mechanizing and automating many agricultural operations. In many cases better accuracy and reliability are obtained by the use of electrical controls compared to manual operation. Because of their ability to safely and accurately control equipment, use of electrical controls in agriculture is continually increasing.

7.1 OPEN-LOOP AND CLOSED-LOOP SYSTEMS

Control systems can be classified as either open-loop or closed-loop systems. In an open-loop system the controlling device operates independently of the process variable it controls. This type of system is demonstrated in Fig. 7.1.

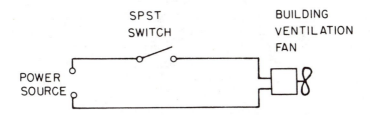

FIG. 7.1. FAN CONTROLLED BY SPST SWITCH, OPEN-LOOP CONTROL SYSTEM

The switch is the controlling device and the fan the controlled device. The temperature in the building being ventilated may be the process variable. In the example, when the switch is closed the fan will operate; when the switch is open it does not operate. A possible disadvantage of an open-loop system is that an operator is required to open or close the switch. Someone must be available at the appropriate times to start and stop the system.

This disadvantage of the open-loop system can be overcome by the closed-loop system whereby the process variable is constantly monitored and kept at a predetermined value automatically. A simple example of a closed-loop system is shown in Fig. 7.2.

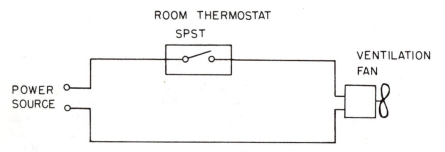

FIG. 7.2.   FAN CONTROLLED BY THERMOSTAT, CLOSED-LOOP CONTROL SYSTEM

The thermostat in Fig. 7.2 is now the controlling device. It may use a bimetallic strip which bends when the temperature of the room changes to open or close the switch. In this way information from the temperature (the process variable) is used to control the switch. This flow of energy or information back from the process to the controller is called feedback.

Note carefully how automatic control is achieved. In the controlling device (controller) the actual value of the process variable is compared to a desired value (the set point). If the difference is large enough the controller acts to reduce the difference. It is important to realize that feedback, flow of energy from the process back to the controller, is an essential element of a closed-loop system. Closed-loop systems lend themselves to automatic operation, thus eliminating the need for human control of each process.

So far we have discussed simple systems where the control is either open (off) or closed (on). In some closed-loop systems feedback can be used to control the rate or speed of the process. For example, the load on a conveyer may be used as feedback to control the speed of the conveyer or the temperature in a building may be used to control the rate of fuel flow into a burner. This type of closed-loop system is called a modulating system.

7.2 SWITCHES AND RELAYS

Simple switches and more complex controls are classified according to the actions the switch can perform. Three designations are used to specify the type of action the switch has:

(1) the number of poles the switch has,
(2) the number of throws the switch has,
(3) whether the switch is normally open or normally closed.

Each movable contact is called a pole. Each stationary contact is called a throw. Thus a single-pole switch has one movable contact, a double-pole switch has two movable contacts operating together, etc. A single-throw switch has only one position in which the poles make contact, a double-throw switch has two positions in which the

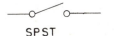

SPST
SINGLE-POLE SINGLE-THROW

SPDT
SINGLE-POLE DOUBLE-THROW

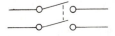

DPST
DOUBLE-POLE SINGLE-THROW

DPDT
DOUBLE-POLE DOUBLE-THROW

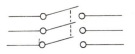

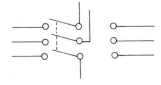

TRIPLE-POLE SINGLE-THROW   TRIPLE-POLE DOUBLE-THROW

FIG. 7.3. BASIC SWITCH DESIGNATIONS

poles make contact, etc. Switches are then designated by their combination of number of poles and number of throws. For example, single-pole single-throw, or double-pole single-throw are switch designations. Some common symbols for the switches of various designations are given in Fig. 7.3.

The designation of normally-open and normally-closed are used to signify the normal position of the switch. If the switch must be actuated to complete the circuit, it is designated normally open (NO). If, on the other hand, the switch must be actuated to open the circuit, it is designated normally closed (NC).

Probably the most common type of control device is the simple switch. Because of their wide use, a few common types of switches will be discussed.

The knife switch consists of hinged metal blades (the movable contacts) and metal clips (the stationary contacts) into which the blades fit. A handle, insulated from the contacts, permits moving the blades. In most applications the knife switch is enclosed in a metal box with the control handle extending outside the box.

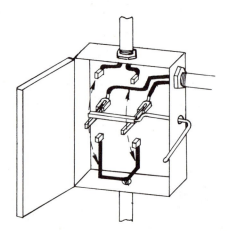

FIG. 7.4.   DPDT KNIFE SWITCH

The toggle-type switch is the type of switch we are accustomed to using in lighting controls. The switch is moved from one position to another by moving an external handle called a toggle. Toggle switches may be single-pole single-throw, double-pole double-throw, single-pole double throw, etc.

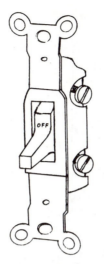

FIG. 7.5.   COMMON SPST TOGGLE SWITCH

In a mercury switch, the switch is closed by liquid mercury completing a circuit between a set of fixed contacts. The liquid mercury
and the contacts are contained within a glass tube. Tilting the tube
makes or breaks the circuit because of the motion of the mercury. A
common application is the auto trunk light. The motion of the trunk
lid tilts the glass tube, thereby activating the switch. A toggle may
also be used to tilt the switch.

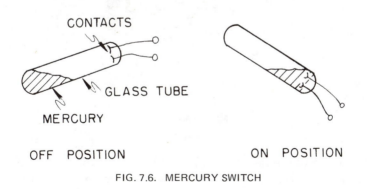

CONTACTS

GLASS TUBE

MERCURY

OFF POSITION   ON POSITION

FIG. 7.6.   MERCURY SWITCH

A number of applications require a switch which will momentarily
open or close a circuit. A push-button switch is commonly used for
this need. If the contacts of the switch are normally closed, pushing
the button opens the switch.

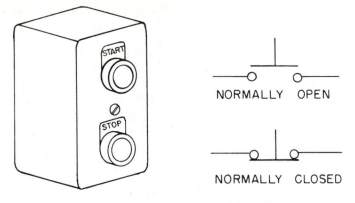

NORMALLY   OPEN

NORMALLY   CLOSED

FIG. 7.7.   START-STOP, "PUSH BUTTON" SWITCH

It remains open as long as the button is held in. When the button is released the switch returns to the closed position. This type of switch is widely used in motor control circuits. Its use will be described more fully when discussing motor controls in Chapter 8.

The snap-action switch is widely used for limit switches. The switch requires very small forces to actuate and quickly snaps from one position to the other. It is often used in a control circuit to sense or limit motion.

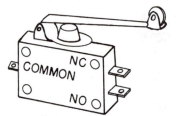

FIG. 7.8.   SIMPLE SNAP-ACTION SWITCH

Electromagnetic relays are very useful in controlling relatively large electrical loads with a second low-power circuit. A small current can operate the relay. The relay then acts as an on-off switch for the main or heavy load circuit.

The basic parts of a relay are the electromagnet and a set of contacts. As shown in Fig. 7.9, the relay is actually part of two separate circuits:

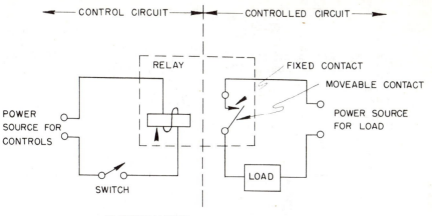

FIG. 7.9.   BASIC RELAY CIRCUIT

(1)   the control circuit — This circuit may be a lower voltage than the load circuit and generally has a very small current flow powering the electromagnets.

(2)   the load circuit — The set of contacts are a switch for the load circuit.

The operation of relays can be explained using Fig. 7.10. For the normally open (NO) relay, the contacts are open until a current through the coil closes them. When the coil is de-energized the spring then reopens the contacts. A normally closed (NC) relay works similarly, except the contacts are opened by current through the coil.

The relay has a distinct advantage over using mechanical switches because a small current in the control circuit can control a relatively large current in the load circuit. Since the control circuit current is small, small wires may be employed. This cuts the expense of wiring for controlling the circuit and allows switches to be placed a considerable distance from the load in remote locations. In addition, heavy currents are not passing through the control circuit; this improves the life of the switches and safety of operation.

Two different types of symbols are commonly used to signify relay contacts. If the contacts are normally closed they appear as ⊣⊬ or ⌐⌐. However, when the contacts are normally open they appear as ⊣⊢ or ⌐⌐ .

Time-delay relays are commonly used to avoid having a number of large loads starting simultaneously or to sequence certain loads. A

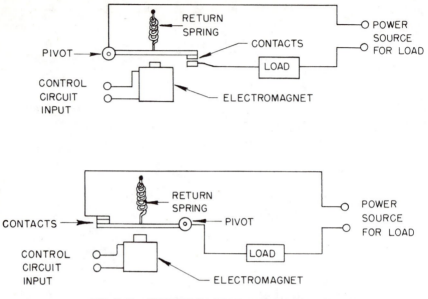

FIG. 7.10. OPERATION OF NO AND NC RELAYS

time-delay relay can be used to open or close a set of contacts a period of time after the control circuit of the time-delay relay has been activated. For a thermal time-delay relay as shown in Fig. 7.11, in the unheated condition the bimetal strip is straight. In this position the movable contact is against the fixed contact on one side. As the strip is heated by current in the control circuit through the

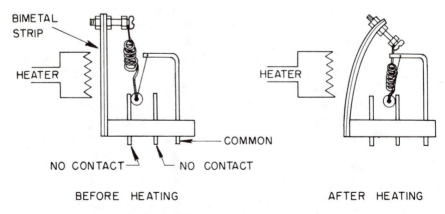

FIG. 7.11. OPERATION OF A TIME-DELAY RELAY

heater, it bends. This causes the movable contact to move across to the other contact. The time-delay is determined by the time required for the strip to bend sufficiently to snap the movable contact across to the other position. Such time-delay relays can be purchased with delays ranging from 3 to 60 seconds. Solid-state electronic time-delays are also available with a wide assortment of delay times.

7.3 SENSING ELEMENTS

One or more of a number of variables including temperature, humidity, pressure, motion, light and time can be sensed and used as a controlled or controlling variable in a control system. The principles of operation of some of the more common devices are discussed in this section.

7.3.1 Temperature Control

Temperature sensing and control devices are commonly called thermostats. One common type of thermostat uses expansion or contraction of a bimetallic strip to sense temperature. A bimetallic strip operates on the principle that as the two metals are heated they expand at different rates. This causes the strip to bend as temperature increases. The movement of the bimetallic strip as it bends opens or closes the contact points.

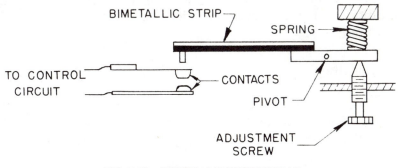

FIG. 7.12. BIMETALLIC THERMOSTAT

A second common type of thermostat is the hydraulic thermostat. The hydraulic thermostat consists of a liquid-filled tube (a capillary tube) mechanically connected to an electrical contact mechanism.

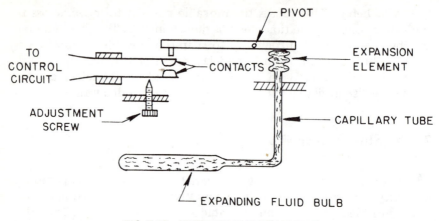

FIG. 7.13. HYDRAULIC THERMOSTAT

Expansion or contraction as the liquid changes temperature opens or closes the contacts.

Thermocouples and thermopiles may also be used in control circuits. As described in section 4.1.3, two dissimilar metals with their ends connected form a thermocouple. If the two junctions are at different temperatures a small voltage proportional to the temperature difference is produced. This voltage can then be amplified by an auxiliary circuit and used as a controlling signal.

Another alternative is to connect a number of thermocouples in series to attain a larger power output. A device which does this is called a thermopile. The output of thermopiles can be used directly if the power requirement for the control circuit is small. A common application of a thermopile is a pilot flame detector controlling a safety valve.

7.3.2. Humdity Control

Humidity sensing devices are commonly called humidistats. They may use human hair or more commonly some other material which responds to changes in moisture or humidity by changing its length. The sensing element is connected mechanically to the contacts such that changes in its length open or close the contacts.

7.3.3. Time Control

Control devices which use time as the sensed variable are widely used in open-loop control systems. Time clocks generally use cams

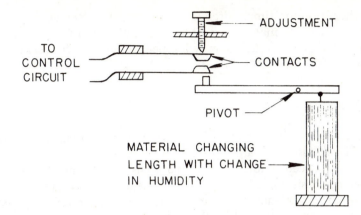

FIG. 7.14.   BASIC OPERATION OF A HUMIDISTAT

controlling snapaction switches to open or close the control circuit at specified time intervals. Time clocks are available with various time intervals for repeating their cycles from every few seconds to every 24 hours or more.

One common type is the 24 hour general purpose time switch. This device makes one revolution in 24 hours and will open and close a switch at intervals set by adjusting cams or removable trippers.

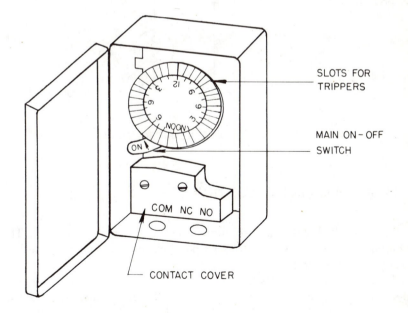

FIG. 7.15.   GENERAL PURPOSE 24 HOUR TIME SWITCH

Another common time switch is a repeating cycle percentage timer. This type of timer may be preset to close the switch for any portion of the timer's cycle. The timer will repeat the on or off cycle as long as power is supplied to it.

7.3.4 Light Control

A photocell senses light and is most often used in lighting applications. Street lights or yard lights can be automatically turned on and off by a photocell. Photocells use materials which when exposed to light change resistance. When light is not falling on the cell its resistance is large and therefore current flow in the circuit is not large enough to activate the relay. When the cell is exposed to light its resistance decreases, the current increases, and the relay is activated.

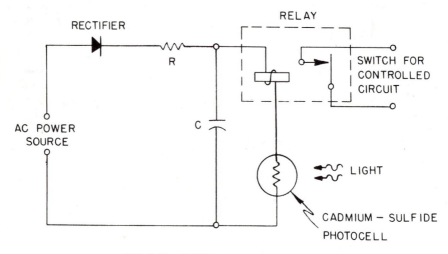

FIG. 7.16. PHOTOCELL CONTROL CIRCUIT

7.4 COMBINATIONS OF CONTROLS

Many control systems use more than one sensing element. The elements can be either in series or parallel, depending on the desired function of the circuit.

If the load is to be activated by a change in any one of a number of elements, the sensing elements are connected in parallel. If the device is to be activated only when all the sensing elements are in the correct position, the sensing elements should be in series.

For example, if you wanted a ventilation fan in a building to operate ten minutes of each hour, or whenever the temperature is above 25°C, or the humidity above 70%, then the control elements would be connected in parallel as in Fig. 7.17. Controls in parallel correspond to "or" situations.

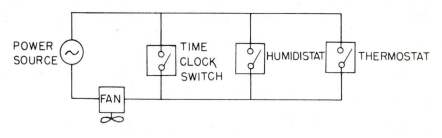

FIG. 7.17.   CONTROLS IN SIMPLE PARALLEL

Controls in series correspond to an "and" case. For example, if you want a pump to operate to fill a tank when the tank is less than half-full, but only at night when the temperature is above freezing, the controls would need to be in series as in Fig. 7.18.

More often combinations of series and parallel elements will be needed when multiple control elements are used.

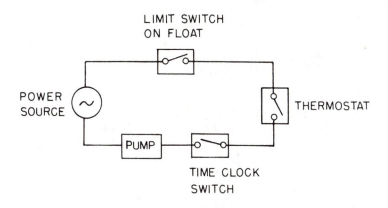

FIG. 7.18.   CONTROLS IN SIMPLE SERIES

Example 7.1.

Combination Series-Parallel Control Elements.

Show schematically a control system for the following situation. A farmer wishes to control his yard light such that it comes on at dusk but turns off at 1 A.M. He would also like to be able to manually turn the light on at any time.

The circuit suggested would use a time clock in series with a photocell. A manual switch is placed in parallel with the other two.

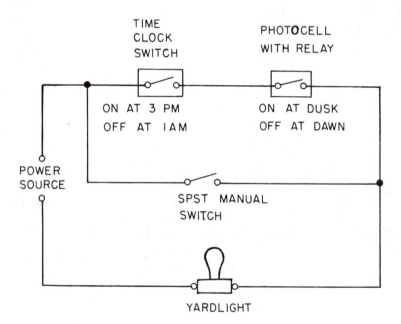

7.5 PLANNING CONTROL SYSTEMS

When planning a control system many decisions as to what variables must be sensed or controlled, what items need to be controlled, and what types of sensors are to be used must be made. A complete working knowledge of the process is necessary to plan the correct controls combined in an effective manner. In addition, controls which are needed for safety reasons must be considered. When automatic controls are used the need for backup manual controls should be considered.

The following example is intended to demonstrate some of these factors for a common control system.

Example 7.2.

Home Furnace Control System.

Consider what control system might be needed for a forced-air LP gas home furnace. A schematic of the basic system is shown below.

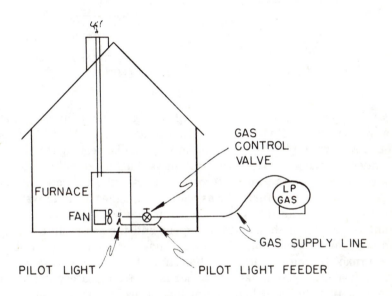

The variable which we want to sense and control in this situation is the temperature in the house. To do that we will need to control the flow of gas into the house and the furnace fan.

The most common system is to use a thermostat in the house to control the flow of gas into the furnace. This would require a normally open thermostat which closes on temperature drop. When the temperature drops below the set point of the thermostat it closes, which in turn activates a NC solenoid valve allowing gas to flow. It takes some time for the furnace to warm up, therefore, to avoid blowing cold air around the house the fan is controlled by a second thermostat in the furnace. A normal open thermostat which closes on temperature rise is needed for this task.

The basic control circuit would appear as:

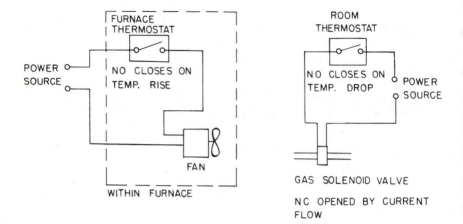

Because the power requirement of the gas valve is low, the valve control circuit is often run on a low voltage (12 or 24 volt) system. This means wiring to the room thermostat can be very light wire.

Under normal conditions the system as diagramed will operate. As the room cools the room thermostat activates the gas valve. After a short time when the furnace has warmed up, the fan is activated by the furnace thermostat. When the room has warmed up to the set point of the room thermostat, it opens and the gas is shut off. The fan continues to run until the furnace has cooled to the set point of the furnace thermostat and it is then shut off.

However, this system has two major faults from a safety standpoint. The first is if the pilot light is not lit, the basement will rapidly fill with gas, creating an obvious hazard. Secondly, if the pilot does light the gas but the fan does not come on, the furnace is in danger of being greatly overheated and will likely become warm enough to start a fire.

The first difficulty can be overcome by adding a thermopile controlled safety valve ahead of the main valve and pilot light feeder. When the pilot light is lit, the thermopile will produce enough current to hold the safety valve open. If the pilot is out, the valve will spring closed.

The second difficulty, the non-starting fan, can be overcome by placing a second thermostat in the furnace. This normally closed thermostat is placed in series with the room thermostat. Its set point is such that if the furnace temperature starts to become excessive it opens the gas valve circuit so that the gas is then shut off.

The final control schematic would appear as:

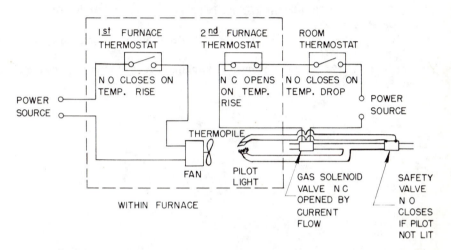

Commonly the two furnace thermostats are purchased as one device. A manual override for the safety valve is needed to light the pilot light of this system.

This system will perform the task required, control the temperature, in a safe and convenient manner. Other additional features can be added to such a control system, but this demonstrates the type of planning and considerations which are involved in developing an electrical control system.

EXERCISES

1. Design a control system for an electrical water heater. Be sure the heating elements will not be powered if no water is in the water heater and that the system has a high water temperature safety control.
2. Draw a control circuit diagram for a system which will ventilate a building when the temperature is above 30°C or the humidity is above 60% in the building.
3. Add to the system in problem two a heater controlled to add heat when the building is being ventilated and the temperature is below 10°C.

REFERENCES

BUTCHBAKER, A. F. 1977. Electricity and Electronics for Agriculture. Iowa St. Univ. Press, Ames, Ia.

HERRICK, C. N. 1975. Electric Wiring: Principles and Practices. Prentice-Hall, Englewood Cliffs, N.J.

MARCUS, A. 1966. Automated Industrial Controls. Prentice-Hall, Englewood Cliffs, N.J.

MERKEL, J. A. 1974. Basic Engineering Principles. AVI Publishing, Westport, Conn.

8

Electric Motors

8.1 ADVANTAGES OF ELECTRIC MOTORS

One of the principal advantages of electrical energy is the ease by which it can be converted to mechanical energy. Over 60% of the electrical energy generated in the U.S. is used by electric motors, according to the Department of Energy. The electric motor is an efficient means of converting electrical energy into mechanical energy. As shown below, efficiency of an electric motor surpassses that of both gasoline and diesel engines.

Approximate Efficiency
Electric Motor	50–70%
Gasoline Engine	25%
Diesel Engine	40%

Electric motors have many advantages over other ways of producing mechanical energy, including:

(1) low initial cost
(2) relatively inexpensive to operate
(3) easy to start
(4) capable of starting a reasonable load
(5) can be automatically and remotely controlled
(6) capable of withstanding temporary overloads
(7) long life, many motors are designed for 35,000 hours of operation

(8) compact
(9) simple to operate
(10) low noise level
(11) no exhaust fumes
(12) minimum of safety hazards.

To make use of these advantages we need to understand the basic principles of how an electric motor converts electrical energy to mechanical energy, the characteristics of various types of motors, how the characteristics are measured and how motors are controlled.

Electric motors are classified in several ways. One classification is by the type of electrical service they require: single-phase alternating current, three-phase alternating current or direct current. Other classification systems are based on such items as type of starting mechanism, rotor style, frame style, application and power output.

8.2 AC MOTOR PRINCIPLES

The vast majority of electrical motors used in homes and on farms are alternating current motors. To understand the principles of operation of a simple AC motor, we must briefly review three basic electrical principles discussed in previous chapters: properties of electromagnets, electromagnetic induction and alternating current.

An electromagnet can be produced by winding insulated wire around a soft iron core. When current passes through the coil of wire, a magnetic field is produced with a north (N) pole at one end of the iron core and south (S) pole at the other. The orientation of the N and S poles is dependent on the direction of current flow and changes each time the current changes direction. It is important to remember that the electromagnet produces a magnetic field only when current is flowing in the coil.

Induction is the phenomenon by which a current is induced in a conductor as it passes through a magnetic field or the field varies around the conductor. As discussed in Chapter 4, the direction of the current flow depends on the direction of the wire movement and the orientation of the magnetic field. The magnitude of the induced voltage is controlled by 1) the strength of the magnetic field, 2) the rate at which the flux lines of the magnetic field are being cut by the conductor, and 3) the number of conductors cutting across the magnetic field.

The third principle to be reviewed is that of alternating current. Current which periodically changes its direction of flow is alternating current. Current in the U.S. is generally generated at 60 hertz, or

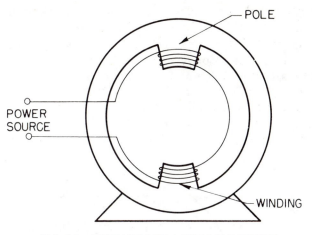

FIG. 8.1. SCHEMATIC OF A TWO-POLE STATOR

cycles per second, meaning the current changes direction of flow 120 times each second.

Combining the principles reviewed, we can show how an electric motor operates. An electric motor is designed with a stationary part called a *stator* and a rotating part called a *rotor*.

The stationary housing contains pairs of slotted cores made up of thin sections of soft iron. The cores are wound with insulated copper wire to form one or more pairs of definite magnetic poles. This stationary part of the motor is the stator. The stator windings will be connected to an AC source to form electromagnets.

One common type of rotor, the squirrel cage rotor, derives its name from its resemblance to an exercise cage for pet squirrels. For

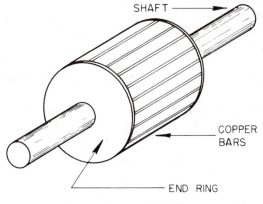

FIG. 8.2. SQUIRREL-CAGE ROTOR

a squirrel cage rotor, a cylinder made up of thin sections of a special soft steel has slots cut in the surface. Bare copper, brass or aluminum bars are mounted in the slots. The bars are short circuited at each end by end rings. The rotor must be carefully balanced on a central shaft. The shaft extends beyond its support bearings at one or both ends to provide for pulleys or other drive mechanisms. Another type of rotor, the wound rotor, will be discussed later.

Assume we insert a simplified rotor into the stator in the position shown in Fig. 8.3.

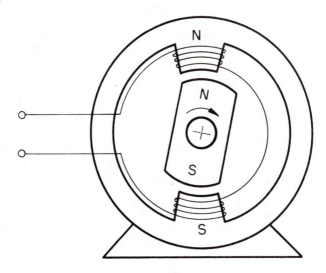

FIG. 8.3. SIMPLE AC MOTOR, POSITION 1

If the poles of the electromagnet (stator) are as shown, the north pole will induce a north pole in the upper portion of the rotor. Likewise the south pole will induce a south in the lower portion of the rotor. Because like poles tend to repel each other, the rotor will rotate clockwise. When the rotor arrives at the horizontal position (Fig. 8.4), the unlike poles will tend to attract, drawing the rotor further around.

If, as the rotor again approaches a vertical position (180° rotation from start), the polarity of the stator poles are reversed, the rotor will continue to be rotated in the same direction.

If the stator is connected to an AC source, the orientation of the electromagnetic poles will continue to alternate. As the rotor continues to spin, theoretically it will adjust itself to the frequency of the source. For a 60 hertz source this would mean a rotational speed

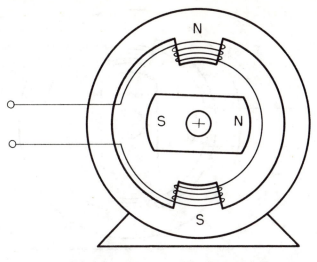

FIG. 8.4.   SIMPLE AC MOTOR, POSITION 2

of 60 revolutions per second or 3600 revolutions per minute for the simple two-pole motor. However, in practice, the speed is less than the theoretical speed due to slip caused by the induced currents in the rotor. Usually under no load, a motor runs 4 to 5% slower than the theoretical speed.

As more sets of poles are added to the stator, the speed of the motor is reduced. The speed of a motor can be expressed as a function of the number of poles and the frequency of the source as,

$$\frac{\text{Revolutions}}{\text{per Minute}} = \frac{\text{Frequency of Source}}{(\text{Number of Poles}/2)} \times \frac{60 \text{ sec.}}{1 \text{ min.}}$$

8.3 SINGLE-PHASE MOTORS

The most common type of motor used in the home, on the farm, and in light industry is the single-phase, alternating current motor. If the rotor of the simple motor described earlier were to stop with the rotor in the alignment shown in Fig. 8.5, there would be no force to start the rotor turning since the poles are in line. Therefore, all single-phase motors require some type of starting mechanism. The starting torque available and starting current requirements will vary with the type of mechanism used. Often single-phase motors are classed by their type of starting mechanism.

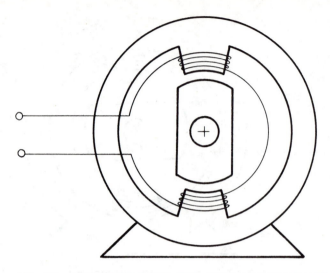

FIG. 8.5. NON-START POINT FOR SIMPLE SINGLE-PHASE
MOTOR.

During the starting period, a current of a magnitude 2 to 7 times
larger than the full-load current is required. The magnitude of the
surge will depend on the motor type and design as well as the load to
be started.

Seven general types of single-phase, alternating current motors
commonly found on farms are:

(1) Split-Phase (SP)
(2) Capacitor
    (a) Capacitor Start (CS-IR, Capacitor Start-Induction Run)
    (b) Two-Value Capacitor (CS-CR, Capacitor Start-Capacitor
        Run)
    (c) Permanent-Split Capacitor (PSC)
(3) Wound Rotor
    (a) Repulsion-Start (RS)
    (b) Repulsion-Induction (RI)
    (c) Repulsion (R)
(4) Shaded Pole
(5) Universal or Series (UNIV)
(6) Synchronous
(7) Soft Start (SS)

The type of motor to select largely depends on the starting requirements of the equipment to be driven and the maximum current that

may be drawn from the single-phase power service. Selection of motors will be discussed more fully in section 8.5. The following sections will briefly discuss the design and operating characteristics of the various types of single-phase motors. Table 8.1 summarizes some of the important characteristics of each single-phase motor type.

8.3.1 Split-Phase Motors (SP)

Split-phase motors are inexpensive and widely used for fractional (less than one) horsepower applications. The mechanism used to start a split-phase motor is a second starting or auxiliary winding connected in parallel with the main stator winding. This auxiliary winding is made of smaller wire and with fewer turns than the main windings. Due to the difference in the size of wire and number of turns, the current and magnetic field reach a maximum in the auxiliary winding before the main winding peaks. The rotor is first acted upon by the auxiliary windings and then by the main windings. The action of the two sets of windings starts the motor. The technique of splitting the single-phase current into two phases yields the name split-phase motor.

Once the motor reaches approximately 3/4 of full speed, the auxiliary windings are deactivated, usually by a centrifugal switch. The auxiliary windings of this type of motor are not designed to operate for extended periods of time. If the motor does not come up to speed, or for some other reason the auxiliary windings are not deactivated, heat build-up will likely damage or "burn out" the auxiliary windings. Direction of rotation of this motor type can be changed by reversing the line connections to the auxiliary windings.

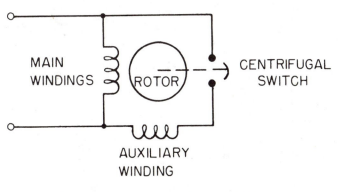

FIG. 8.6.   SPLIT-PHASE MOTOR SCHEMATIC

TABLE 8.1. TYPES OF SINGLE PHASE MOTORS AND THEIR CHARACTERISTICS

Type	Horsepower Ranges	Load-Starting Ability	Starting Current	Characteristics	Electrically Reversible	Typical Uses
Split-phase	1/20 to 1/2	Easy starting loads. Develops 150 percent of full-load torque.	High; five to seven times full-load current.	Inexpensive, simple construction. Small for a given motor power. Nearly constant speed with a varying load.	Yes.	Fans, centrifugal pumps; loads that increase as speed increases.
Capacitor-start	1/8 to 10	Hard starting loads. Develops 350 to 400 percent of full-load torque.	Medium, three to six times full-load current.	Simple construction, long service. Good general-purpose motor suitable for most jobs. Nearly constant speed with varying load.	Yes.	Compressors, grain augers, conveyors, pumps. Specifically designed capacitor motors are suitable for silo unloaders and barn cleaners.
Two-value capacitor	2 to 20	Hard starting loads. Develops 350 to 450 percent of full-load torque.	Medium, three to five times full-load current.	Simple construction, long service, with minimum maintenance. Requires more space to accommodate larger capacitor. Low line current. Nearly constant speed with a varying load.	Yes.	Conveyors, barn cleaners, elevators, silo unloaders.
Permanent-split capacitor	1/20 to 1	Easy starting loads. Develops 150 percent of full-load torque.	Low, two to four times full-load current.	Inexpensive, simple construction. Has no start winding switch. Speed can be reduced by lowering the voltage for fans and similar units.	Yes.	Fans and blowers.

Type	hp range	Starting characteristics	Starting current	Operating characteristics	Reversible	Applications
Shaded pole	1/250 to 1/2	Easy starting loads.	Medium	Inexpensive, moderate efficiency, for light duty.	No.	Small blowers, fans, small appliances.
Wound-rotor (Repulsion)	1/6 to 10	Very hard starting loads. Develops 350 to 400 percent of full-load torque.	Low, two to four times full-load current.	Larger than equivalent size split-phase or capacitor motor. Running current varies only slightly with load.	No. Reversed by brush ring re-adjustment	Conveyors, drag burr mills, deep-well pumps, hoists, silo unloaders, bucket elevators.
Universal or series	1/150 to 2	Hard starting loads. Develops 350 to 450 percent of full-load torque.	High.	High speed, small size for a given horsepower. Usually directly connected to load. Speed changes with load variations.	Yes, some types.	Portable tools, kitchen appliances.
Synchronous	Very small, fractional	N/A[1]	N/A	Constant speed	N/A	Clocks, timers,
Soft-start	10 to 75	Easy starting loads.	Low, 1.5 to 2 times full-load current.	Excellent for large loads requiring low starting torque.	Yes.	Crop driers, forage blowers, irrigation pumps, manure agitators.

Source: Soderholm and Puckett (1974).
[1] N/A = not applicable.

The smaller-size wire used for the auxiliary windings has the advantage of a small space requirement. However, the small wire limits the starting current and starting torque of the motor. Split-phase motors are only suitable for handling easy starting loads such as ventilation fans. They are rarely used for motors larger than one-half horsepower because of their relatively high starting current factor. Generally split-phase motors are limited to low starting torque applications where low cost is more important than high starting currents.

8.3.2 Capacitor Motors

Simple capacitor-start motors are nearly the same as split-phase motors, except that a capacitor is connected in series with the auxiliary windings.

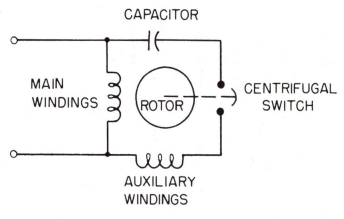

FIG. 8.7.   CAPACITOR-START MOTOR (CS–IR)

The capacitor creates a larger phase shift between the starting and running windings. Two reasons the capacitor improves starting characteristics are:

(1) the increased split in the single-phase currents creates a wider time interval between the two magnetic field peaks.
(2) it allows more copper in the auxiliary winding and thus reduces the starting current requirements.

The simple capacitor start motor described by Fig. 8.7 has approximately twice the starting torque with about one-third less starting current than a split-phase motor. The capacitor-start motor is also reversible by reversing the auxiliary winding leads.

Two other types of capacitor motors are available. These two types differ from the standard capacitor-start motor in that for both the auxiliary winding remains in the circuit at all times. This implies the wire used for the auxiliary winding must be as heavy as the main windings to withstand the heat build-up due to the continuous current flow.

Two-value capacitor (CS-CR, capacitor start-capacitor run) motors are similar to capacitor-start motors for starting. However during

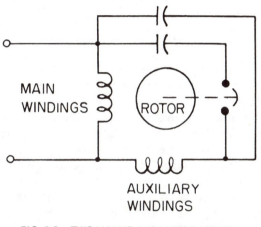

FIG. 8.8.   TWO-VALUE CAPACITOR MOTOR

running a small capacitor remains in series with the auxiliary windings. This capacitor gives greater efficiency by lowering the line current required by the motor. CS-CR motors have slightly higher starting torque than CS-IR motors and can therefore handle more difficult starting loads. Starting current requirements for the two types are about the same.

Permanent-split capacitor (PSC) motors are similar to CS-CR motors except the same value of capacitance is used for both starting and running. This has the advantage of eliminating a centrifugal switch, however the starting torques for these motors are much lower than for other capacitor motors.

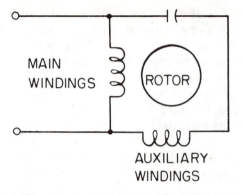

FIG. 8.9. PERMANENT-SPLIT CAPACITOR MOTOR

8.3.3 Wound-Rotor Motors

These motors get their name from the fact that their rotors are made-up of wire windings connected to a commutator ring and brushes much like a generator armature. The brushes are short-circuited and shifted to give the effect of a second stator winding. The stator winding current induces a current in the rotor windings. The fields from the two currents oppose each other and thus produce a torque. These motors have good starting torque with low starting currents and are therefore commonly used for heavy starting loads. These motors are more expensive than split-phase or capacitor motors and also require more maintenance because of brush and commutator wear. Two types of wound-rotor motors used for agricultural applications are discussed in more detail in the following paragraphs.

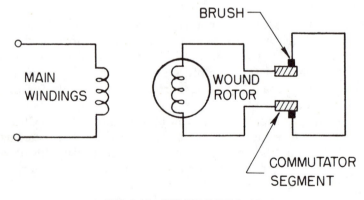

FIG. 8.10. WOUND-ROTOR MOTOR

Repulsion-start induction-run (RS) motors start as a repulsion motor but switch to operate as an induction motor. At a predetermined speed, all the rotor windings are short circuited to give the equivalent of a squirrel-cage winding. The repulsion-start induction-run motors are the most common type of wound-rotor motors.

Repulsion motors (R) is a term often used for all wound-rotor motors. However, a true repulsion motor is a type in which the brushes short-circuit selected windings on the commutator in such a manner that the magnetic axis of the rotor is shifted from the magnetic axis of the stator. Speed of this type of motor is controlled by the load. This type of motor is sometimes referred to as a variable-speed motor.

8.3.4 Shaded-Pole Motors

Shaded-pole motors are simply constructed, low-cost motors for loads with low-starting torque requirements. Instead of an auxiliary winding, shaded-pole motors have a continuous solid copper loop around a small portion of each pole. The shading loop causes a reaction which gives the motor some starting torque. Low efficiency as well as low starting torque and poor power factor limit the use of this type of motor to light loads such as small fans.

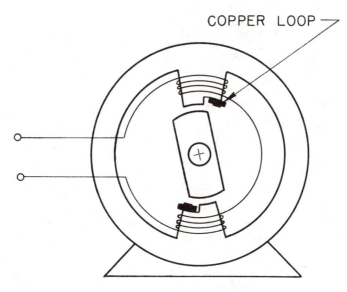

FIG. 8.11. SHADED-POLE MOTOR.

8.3.5 Universal or Series Motors (UNIV)

The universal or series motor gets its designation from having the stator and rotor windings in series. This type of motor will operate on either AC or DC power sources. It is usually used as a special purpose motor. Often it is built into portable equipment such as drills, grinders, sanders, vacuum cleaners and food mixers. The advantages of this type of motor include high power-to-size ratio and

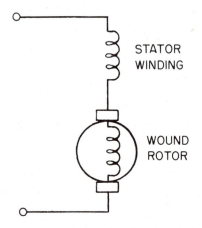

STATOR WINDING

WOUND ROTOR

FIG. 8.12.   UNIVERSAL OR SERIES MOTOR

rapid acceleration. This type of motor does not operate at a fixed speed, but rather runs as fast as the load allows. A good example of this characteristic is an electric hand drill where the motor slows as the load increases. With no load on the drill, the friction of the bearings and motor limit the speed.

8.3.6 Soft-Start Motors (SS)

In large integral (greater than one) horsepower motors, starting current requirements may limit the size of a motor which may be used on a single-phase line. In many cases single-phase motors over five or seven and one-half horsepower will need to be of the soft-start type. Soft-start motors limit the starting current to as low as one and one-half to two times the full-load current. By reducing the starting current surge, higher horsepower motors can be used on the single-phase line without adversely affecting other loads on the line. However, the reduced starting current also reduces the available starting torque to 50 to 90% of full-load torque. Therefore this type of motor is matched to easy starting loads such as drier fans, forage blowers and irrigation pumps.

The lower starting currents are often obtained by switching the orientation of motor windings. Two windings placed in series for starting are switched to parallel for running as the motor accelerates.

8.4 THREE-PHASE MOTORS

A three-phase motor has a set of stator windings for each of the phases. Three-phase motor windings may be either wye or delta connected (Fig. 8.13). For balanced phase voltages, both types are similar in performance.

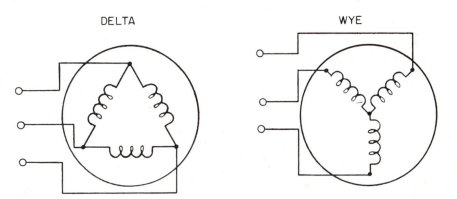

FIG. 8.13.   THREE-PHASE WYE AND DELTA MOTORS

Having three-phase AC power permits a simple, low-cost design. With each of the three windings, each 120° apart, there are no positions where torque is not produced to turn the rotor, as there are for single-phase motors. Therefore three-phase motors do not require a starting mechanism; they are inherently self starting. This eliminates many expense and maintenance problems.

Three-phase motors are common in horsepowers from one-half to four hundred. Starting torque is generally high with low-to-moderate starting currents, three to four times full-load current. Some typical uses of three-phase motors include crop dryer fans, irrigation pumps, elevators and conveyers. Use of three-phase motors is often limited by the availability of three-phase power. More discussion of the use of three-phase motors was given in Section 5.5.

8.5 MOTOR RATINGS AND SELECTION

Successful motor selection primarily entails choosing a motor that will meet load requirements without exceeding temperature and torque limitations of that motor. The first step in motor selection is to determine the load characteristics, such as power or torque requirements, speed and duty cycle.

Starting and running torques must both be considered. Starting torque requirements vary by the type of load from a small percentage of full-load, as for fans, to several times full-load, as for barn cleaners or loaded augers. At all times from start to full speed, the torque supplied by the motor must be more than that required by the load. The greater the excess torque, the more rapid the acceleration. By examining motor performance curves, we can determine if a motor has enough torque to start the load, accelerate to full speed and handle the maximum overload which may occur.

Basic motor performance is described by a speed vs. torque curve. The curve shown in Fig. 8.14 is for a general-purpose, squirrel-cage induction motor. The figure shows how output varies as speed increases from zero to synchronous speed.

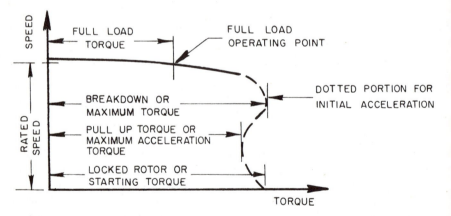

FIG. 8.14. SPEED VS. TORQUE FOR A GENERAL-PURPOSE SQUIRREL-CAGE MOTOR

Several locations on the speed vs. torque curve have been given special names because of their significance in matching motor and load characteristics. They are:

locked-rotor torque — motor torque at zero speed or the maximum torque available to start the load.

pull-up torque — lowest value of torque produced by the motor between zero and full-load speed.

full-load torque — torque output at rated speed.

break-down torque — maximum value of torque produced by the motor during overload without stalling.

In addition acceleration torque is defined as the difference between motor torque produced and torque required by the load at any given time during acceleration. This is the torque available to further accelerate the load.

Once the required motor torque characteristics have been met, several other factors about the motor design need to be considered. They include starting current requirements, temperature rating, duty, operating environment and service factor.

In cases where the maximum starting current which the motor draws may be a limit, the designation of the starting current for the motor is helpful. A motor code, designated by a letter on the nameplate, indicates the starting current required. Table 8.2 shows some of the common letter designations. The higher the locked-rotor kilowatt-ampere rating the higher the starting current surge will be.

TABLE 8.2. MOTOR CODE LETERS, APPLIED TO MOTORS STARTING ON FULL VOLTAGE

Code Letter	Locked Rotor KVa per hp
F	5.0 to 5.6
G	5.6 to 6.3
H	6.3 to 7.1
J	7.1 to 8.0
K	8.0 to 9.0
L	9.0 to 10.0

Source: Anon. 1972. Standards MG 1-10.36. Motor Code Letters. NEMA, N.Y.

Example 8.1.

Locked-Rotor Current Calculation.

Calculate the approximate locked-rotor current for a 1/2 horsepower, 240 volt motor with an H motor code.

H motor code implies 6.3 to 7.1 KVa/hp

$$6,300 \frac{Va}{hp} \times 1/2 \, hp \times \frac{1}{240 \, V} = 13.1 \, amp$$

$$7,100 \frac{Va}{hp} \times 1/2 \, hp \times \frac{1}{240 \, V} = 14.8 \, amp$$

Locked-rotor current would be between 13.1 and 14.8 amperes.

If the motor torque and current requirements fit the application, motor temperature rating should be considered. Both bearing and insulation life are reduced as the operating temperature of the motor increases. However, if the motor temperature is maintained at a safe level for the insulation, bearing requirements are generally also met.

Four insulation systems are common for small motors. They are classed by the maximum temperature any spot in the motor can tolerate.

Insulation System	Maximum Hot Spot Continuous Temperature
Class A	105°C (221°F)
Class B	130°C (266°F)
Class F	155°C (311°F)
Class H	180°C (356°F)

Nameplate data most often gives the permissible temperature rise above the ambient air or the maximum ambient temperature that will keep hot spot temperature within the specified limit. Normal maximum ambient temperature for motor operation is 40°C (104°F) for most motor ratings. If the motor is to operate in an environment above that temperature, special considerations must be made for cooling the motor.

Manufacturers often classify motors as continuous duty or intermittent duty. Motor duty refers to how frequently the motor is started and for how long it will run each time it is started. Continuous duty is defined as the type of service in which the motor is operated at or near full load for more than 60 minutes at a time. This would be the common situation for many loads. Intermittent duty is the type of service in which the load is only on for 10, 20 or 30 minutes at a time with a rest or cooling period between operations. Some examples of this type of load which may be serviced by an intermittent duty motor include automatic air compressors, furnace fans, refrigerators and domestic water pumps.

Most motors are designed for continuous duty. The reason for making intermittent duty motors is a matter of cost. Heat dissipation is not as critical on an intermittent duty motor; therefore, some components can be constructed less expensively. For a person purchasing only one or two motors, the difference in cost may not be enough to compensate for the reduced adaptability and flexibility of the motor. If the operating period is overextended for any reason, the intermittent duty motor may overheat and burn out prematurely, thereby negating any initial cost savings.

Ideally it is best to keep a motor as open as possible to allow excess heat from the windings to be dissipated. However, when operating in an environment which may be harmful to the motor because of water or dust, the motor needs an enclosure which will protect it. Three common types of enclosures are drip proof, splash proof and totally enclosed.

A drip-proof enclosure, often called the open-type enclosure, allows for easy movement of air through slots in the end shields. This is highly desirable for motors operating in clean air and where water is kept from entering the motor.

Splash-proof enclosures are designed to keep water from splashing into the motor. This type of enclosure protects the motor from falling water and also from water striking from the side, unless from a very low angle. Splash-proof enclosures are very common around the farm. Splash-proof motors cost approximately 15 to 20% more than drip-proof motors of the same horsepower.

Totally enclosed motors provide no openings for circulation of outside air through the motor. A fan on the rotor circulates air within the housing. One advantage of totally enclosed motors is that they are not affected by dusty conditions. Dust can accumulate in partially enclosed motors and interfere with proper cooling. Totally enclosed motors should be used in dusty environments such as feed mills or mixers. Totally enclosed motors cost 20 to 40% more than drip-proof motors but may reduce repair bills in the long run.

Other types of enclosures are available which protect against corrosive environments, allow the motor to be submerged in a liquid, or protect against ignition of explosive gases.

There are two main types of bearings used for motors, sleeve bearings (Fig. 8.15) and ball bearings (Fig. 8.16). The choice of which type to use depends mainly on the method and frequency of lubrication and the mounting orientation.

The sleeve bearing consists of a brass or bronze collar in which the shaft rotates. Sleeve bearings generally require more frequent lubrication than ball bearings and are not well adapted to mounting positions where the motor shaft is not nearly horizontal.

Ball bearings consist of steel balls that roll in a special cage around the shaft. The ball bearing has less friction and requires less frequent lubrication.

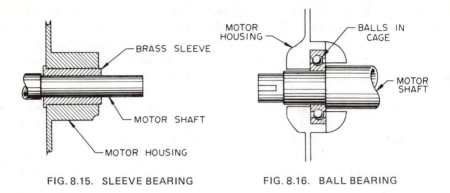

FIG. 8.15.   SLEEVE BEARING   FIG. 8.16.   BALL BEARING

Standard size frames and shaft heights have been established by the National Electrical Manufacturers Association (NEMA) for integral horsepower motors manufactured since 1964. Standardization allows interchangability between motors from different manufacturers. A NEMA frame designation should appear on the motor nameplate. Shaft height in inches for integral horsepower motors may be obtained by dividing the first two numbers of the frame size by four. Shaft height in inches for fractional horsepower motors may be obtained by dividing the frame size by 16.

Example 8.2.

Shaft Height Calculated.

Determine shaft height for the following two motors:

(a)   2 hp, frame size 180
(b)   1/2 hp, frame size 40

Solution:

(a) $\dfrac{18}{4} = 4.5$ in.

(b) $\dfrac{40}{16} = 2.5$ in.

The motor nameplate carries a good deal of the essential information about the motor. A typical nameplate is shown in Fig. 8.17.

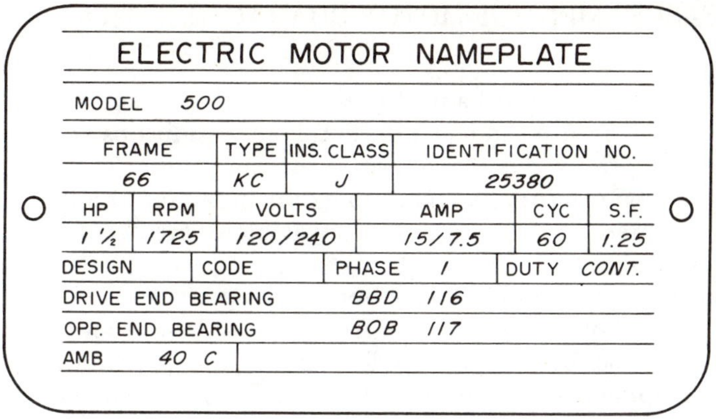

ELECTRIC MOTOR NAMEPLATE

MODEL   *500*

FRAME	TYPE	INS. CLASS	IDENTIFICATION NO.		
66	*KC*	*J*	*25380*		

HP	RPM	VOLTS	AMP	CYC	S.F.
1 ½	*1725*	*120/240*	*15/7.5*	*60*	*1.25*

DESIGN	CODE	PHASE *1*	DUTY *CONT.*

DRIVE END BEARING   *BBD   116*

OPP. END BEARING   *BOB   117*

AMB   *40 C*

FIG. 8.17. TYPICAL ELECTRIC MOTOR NAMEPLATE

The information generally found on the nameplate includes:

(1) Name of the manufacturer
(2) Frame designation — the NEMA designation for frame design
(3) Horsepower — full-load horsepower rating
(4) Motor Code — letter designating starting current requirement
(5) Cycles or Hertz — frequency of the source to be used
(6) Phase — number of phases of the source (one, single-phase, three, three-phase)
(7) Revolutions per minute — speed of the motor at full-load
(8) Voltage — voltage or voltages at which the motor is designed to operate
(9) Thermal protection — indicates if built-in overload protection is provided
(10) Amperes — rated current at full load
(11) Time — duty rating, continuous or intermittent
(12) Ambient temperature or temperature rise — maximum temperature at which the motor should be operated, or temperature rise of the motor above ambient at full-load
(13) Service factor — the amount of overload the motor can tolerate continuously at rated voltage and frequency
(14) Insulation class — a designation of insulation class generally used only for rewinding
(15) Identification of bearings — for replacement of bearings
(16) Power factor — power factor at full load appears on some recently manufactured motors.

8.6 MEASUREMENT OF MOTOR CHARACTERISTICS

By measuring the mechanical output of a motor and the electrical input to the motor under different load conditions, data can be developed to show the characteristics of the motor. The mechanical output can be measured either with a dynamometer or a Prony brake test. A voltmeter, an ammeter, and a wattmeter are necessary to measure electrical input.

A schematic of a simple Prony brake test apparatus is shown in Fig. 8.18. The Prony brake applies a friction load to the motor shaft by means of wood blocks, a flexible band or another friction surface. With this apparatus the torque output of the motor can be measured and controlled.

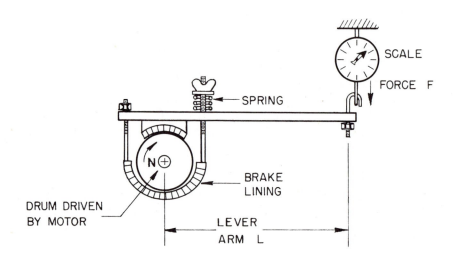

FIG. 8.18.   SIMPLE PRONY BRAKE APPARATUS

From the Prony brake test, torque and horsepower can be calculated. Torque is the lever arm L times the force F. Power or work per unit time can be calculated from the torque and revolutionary speed as,

$$POWER = \frac{WORK}{TIME} = \frac{FORCE \times DISTANCE}{TIME} = F \times 2\pi L \times N$$

In units of horsepower, output power is expressed as,

$$POWER = \frac{2\pi FLN}{33,000} \, hp$$

where   F = force in pounds
L = lever arm in feet
N = rotational speed of shaft in revolutions per minute (RPM)

$$33,000 = \frac{1 \text{ ft-lb/min}}{1 \text{ hp}}$$

In units of watts, output power can be expressed as,

$$POWER = 2\pi FLN \text{ watts}$$

where   F = force in Newtons
L = lever arm in meters
N = rotational speed of shaft in hertz or cycles per second

If instrumentation such as that shown in Fig. 8.19 is used in conjunction with the Prony brake, efficiency and power factor can also be calculated at any load condition. Motor efficiency is the ratio of output power as measured by the Prony brake to the input power as measured by the wattmeter.

$$\% \text{ EFFICIENCY} = \frac{POWER \; OUT}{POWER \; IN} \times 100\%$$

Power factor is the ratio of true to apparent power or,

$$POWER \; FACTOR = \frac{Wattmeter \; Reading}{Volts \times Amperes}$$

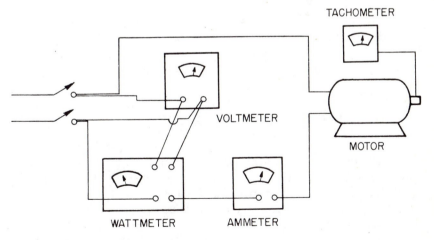

FIG. 8.19. ELECTRICAL INSTRUMENTATION FOR MEASUREMENT OF POWER INPUT TO A SINGLE-PHASE MOTOR

Example 8.3.

Motor Characteristic Calculations.

Calculate the horsepower output, efficiency, and power factor from the following data.

Electrical Meter Readings

I_m = 11.4 amp V_m = 110 ^{V} P = 960 W

From Prony Brake

F = 2 lb   L = 1 ft   N = 1720 RPM

$$\text{POWER} = \frac{2\pi \, FLN}{33000} \, hp = \frac{2\pi \times 2 \times 1 \times 1720}{33000} \, hp$$

$$= 0.66 \, hp$$

$$\% \, EFF = \frac{OUTPUT}{INPUT} \times 100\% = \frac{0.66 \, hp \times 746 \, W/hp}{960 \, W} \times 100\%$$

$$= 51\%$$

$$\text{POWER FACTOR} = \frac{WATTS}{VOLTS \times AMPS} = \frac{960 \, W}{110 \, V \times 11.4 \, amp} = 0.76$$

If efficiency and power factor are calculated over a range of motor loads, plots of efficiency and power factor can be developed. Figure 8.20 shows such a plot for a three-phase motor.

8.7 MOTOR PROTECTION AND CONTROL

No matter what type of motor is selected, to operate the motor we must have a control to start and stop the motor, and some type of overload protection. Since overload protection is often built into the control system, these two topics are discussed together.

Motors must be protected against both excessive current and excessive winding temperatures caused by electrical faults, excessive loads on the motor, or low supply voltage. Sustained overcurrent and over-

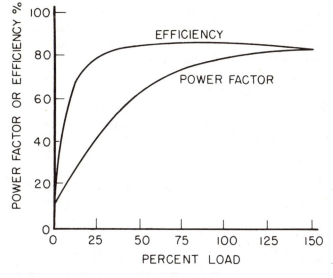

FIG. 8.20 TYPICAL THREE-PHASE MOTOR EFFICIENCY AND POWER FACTOR

heating will cause the insulation in the motor to break down and the motor to "burn up." Branch circuit fuses or circuit breakers may not adequately protect the motor. Branch circuit protection is designed to protect the circuit wires and may be at a rating above that necessary for motor protection.

There are two main types of motor protective equipment available — fuses and thermal-overload devices. Fuses are commonly included with manually operated switches. Thermal-overload devices are used on both manual and electromagnetic controllers.

Time-delay fuses afford both short circuit and overload protection. As described in Section 4.6.3, time-delay fuses can tolerate an overload for a brief period, thereby allowing for starting current surges. Either plug or cartridge-type fuses can be used for motor protection.

Thermal-overload devices with either a bimetallic element, much like a circuit breaker, or an eutectic element, with action much like a normal time-delay fuse, are both available. A thermal-overload switch may be built into the motor itself. The overload device generally opens the line directly on a fractional horsepower motor. With larger motors a relay system may be used.

Thermal-overload devices built into motors or controllers can be either manual-reset or automatic-reset. Manual-reset means a button must be pressed to reset the tripped mechanism. This type is generally recommended for general-purpose motors because the condition causing the overload can be corrected before the motor restarts. Unexpected start-ups which might present safety hazards are avoided. Automatic reset mechanisms automatically attempt to restart the motor after the thermal unit cools.

Overcurrent devices come in a wide range of rated tripping currents. Since the needs of a particular motor may not exactly match a standard, higher ratings than recommended may be necessary. Table 8.3 gives recommended and maximum ratings for overcurrent protection by percentage of full-load current rating.

TABLE 8.3. OVERCURRENT PROTECTION RATING AS % OF FULL-LOAD CURRENT

	Recommended	Maximum
Motors with Service Factor of 1.15	125%	140%
Motors with a marked temperature rise not over 40°C	125%	140%
All other motors	115%	130%

Controls for electric motors can vary from simple on-off toggle switches to complex automatic systems. Some of the simpler controls will be discussed here. Advice on complex motor systems can be obtained from a power supplier or manufacturer.

Manual switches are most often used to control small motors of one-half horsepower or less. These switches are low cost and can be purchased with built-in overcurrent protection. Control switches for electric motors must be able to withstand the high starting current and arcing that occurs when the circuit is opened due to the highly inductive nature of the load. Quick-make, quick-break switches equipped with arc quenchers are used. This type of switch is rated by horsepower and voltage. Regular tumbler-type switches as used for light switches should not be used to control motors. They can withstand the starting current surges, but are not equipped with arc quenchers and therefore usually burn out quickly.

Three basic types of motor circuits using manual control switches are shown in Fig. 8.21. For type 1, the branch circuit protection also supplies the overcurrent protection for an individual motor as the only load on the circuit. For this circuit, the overcurrent protection would need to be sized to protect the motor.

Magnetic motor starters are widely used for controlling motor loads. This type of starter should be used in all motors larger than one horsepower and is an essential element in automatic control

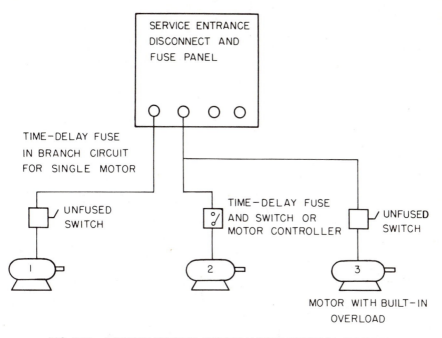

FIG. 8.21.   COMMON MANUAL-SWITCH MOTOR CONTROL CIRCUITS

systems. The difference between a manual switch and a magnetic switch is in how the motor is started. Instead of a manual switch to open or close the circuit, the magnetic switch works with manual or automatic control of a magnetically controlled switch. The operation of the magnetic controller will be described using Fig. 8.22.

The following is the operating sequence for starting and stopping a motor controlled by a magnetic starter:

(1) Pushing the start button (Fig. 8.22) completes the circuit through the coil of the electromagnet. Current flows from L1 through the start switch, the coil, the overload switch and back to L2.

(2) The magnetic action of the electromagnet closes all three sets of contact points and thereby supplies a complete circuit to the motor. In addition, the lower set of closed contacts supplies a circuit parallel to the start switch. Therefore the start switch can be released without deactivating the electromagnet which holds the contacts closed. If the coil is deactivated all three sets of contacts are opened by action of a spring.

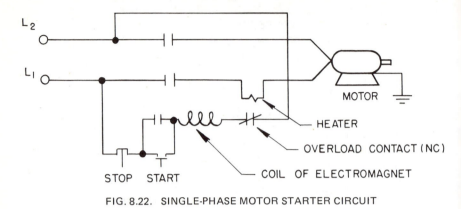

FIG. 8.22. SINGLE-PHASE MOTOR STARTER CIRCUIT

(3) To stop the motor, the coil can be deactivated by momentarily pushing the stop switch. This deactivates the coil and allows all three contacts to open. Once the contacts have opened, the stop button can be released without the motor restarting.

As current passes through the motor circuit, it passes through the heater. Heat is given off, but, under normal conditions, not enough heat is developed to cause the bimetallic overload switch to open. However under an overload situation, the excess heat will cause the overload switch to open. When the overload contacts are opened, the coil loses power and the contact points open, stopping current flow to the motor. To restart the motor, the overload control is reset, and the start button pressed again.

The magnetic motor starter is used as a key element in many control systems. Use of control mechanisms other than Start-Stop switches allows for the development of many automatic control systems. One distinct advantage of the magnetic starter is that current in the control segment of the circuit is very small. The high resistance of the electromagnet coil limits this current. Therefore lighter wires and switches can be used for the control circuit and switches can be located in remote location more readily.

8.8 WIRING FOR MOTOR BRANCH CIRCUITS

For safety purposes the frame of each motor should be connected to the grounding system. If an electrical fault develops in the motor, grounding will prevent hazardous voltages between the motor frame and the earth. By supplying an easy path for current, current due to the fault will trip the overcurrent protection.

Motors perform best at rated voltages, therefore wires must be sized to avoid excessive voltage drop. Not more than a 2% voltage drop is recommended for motor branch circuits.

Tables A.3 and A.4 of Appendix A give full-load currents for single- and three-phase motors. Current values from those tables should be used in the wire selection process unless the motor nameplate current is larger. If the nameplate current is larger, it should be used.

Branch-circuit conductors to an individual motor should be selected to carry 125% of full-load current of the motor with 2% or less voltage drop. The 125% factor allows for starting current surges and a certain degree of overload. When the conductors supply more than one motor on a branch circuit, the wire is sized for a current value of 125% of largest motor current plus 100% of the additional motor currents.

Wire size required to meet voltage drop and current carrying limitations is determined in the same manner as described for feeder wires in Section 5.4 of this text. The following two examples demonstrate the process.

Example 8.4.

Branch Circuit Wire Sizing for Motors, I.

What size copper conductor would be required for a 1/2 horsepower, 120 volt, single-phase motor located 20 meters from the service entrance?

From Table 3, motor full-load current = 9.8 amp

Allowable Voltage Drop = 2% $\times$ 120 V = 2.4 V

$$\text{Allowable Resistance} = \frac{E}{I} = \frac{2.4 \text{ V}}{1.25 \times 9.8 \text{ amp.}} = 0.20 \text{ ohms}$$

Calculating ohms/1000 m

$$\frac{x \text{ ohms}}{1000 \text{ m}} = \frac{0.20 \text{ ohms}}{40 \text{ m}}$$

Allow. R = 5 ohms/1000 m

From Table 1, need #10

Check allowable ampacity, Table A.5 — okay.

Example 8.5.

Branch Circuit Wire Sizing for Motors, II.

Calculate the copper branch circuit conductor size needed for serving the following two motors on the same 240 volt branch circuit if both motors are located 15 m from the service entrance.

   Motor 1 — 3/4 hp, single-phase

   Motor 2 — 1 hp, single-phase

From Table 3. 3/4 hp motor — 6.9 amp full-load

            1 hp motor — 8.0 amp full-load

Total current for calculations = 1.25 × 8.0 amp + 6.9 amp
                              = 16.9 amp

Allowable Voltage Drop = 2% × 240 V = 4.8 V

$$\text{Allowable Resistance} = \frac{E}{I} = \frac{4.8\ V}{16.9\ \text{amp}} = 0.225\ \text{ohms}$$

$$\frac{\text{Allowable Resistance}}{1000\ m} = \frac{0.225 \times 1000}{30} = 7.5\ \frac{\text{ohms}}{1000\ m}$$

From Table A.1 need #12 wire.

Checking Table A.5 for allowable ampacity, #12 okay for 16.9 amp load with any of the insulations listed.

In summary the following measures must be provided for in the wiring system for a motor:

(A) branch-circuit conductors of appropriate size to avoid excess voltage drop

(B) branch-circuit overcurrent protection to protect the con-
ductors of the motor circuit
(C) a means of disconnecting the motor from the electrical supply
(D) motor overcurrent protection to prevent overloading the
motor under running conditions
(E) a controller to start and stop the motor
(F) grounding of the motor frame for safety

EXERCISES

1. An electric motor draws 2238 watts and 12 amperes when operating on a 240 volt 60 hertz source.
Calculate: a. Power factor of motor
              b. Power input to motor
              c. Horespower output of the motor if it is 75%
                  efficient.

2. The following information is found on the nameplate of a motor:

| 3 hp | 60 Hz | 1 phase |
| 120/240 volts | 34/17 amp | 1740 RPM |

a. If the power factor of the motor is 0.80, what is the power
input to the motor at full load?
b. What is the efficiency of the motor at full load?
c. How much does it cost to operate this motor at full load for
100 hours at 3¢ per kWh?

3. What would be the speed of a 120 volt, 6-pole motor on a 50
hertz source?

4. a. How many foot-pounds of work must be done in filling a
30,000 gallon tank with water from a well in which the water
level is 159 feet below the tank? (1 gal. = 8 1/3 lb).
b. What horsepower is needed to fill this tank in two hours
assuming 100% efficiency?
c. If the motor and pump combination is 45% efficient, what
size motor is needed?

5. What size copper wire cable (THW insulation) is needed for the
following motor if located 30 m from the service entrance?

| 3hp | 240 V | single-phase |

6. What size wire is needed for a 5 horsepower, single-phase motor located 12 m from the source? Assume aluminum, UF insulation and cable underground.

REFERENCES

BUTCHBAKER, A. F. 1977. Electricity and Electronics for Agriculture. Iowa St. Univ. Press, Ames, Ia.

LEONARD, M. G., CURRY, D. T., KROUSE, J. K., and TESCHLER, L. E. 1978. Mach. Des. *50* (11) (Electrical and Electronics Reference Issue).

MERKEL, J. A. 1974. Basic Engineering Principles, AVI Publishing, Westport, Conn.

PARADY, W. A. and TURNER, J. H. 1978. Electric Motors. AAVIM, Athens, Ga.

SODERHOLM, L. H. and PUCKETT, H. B. 1974. Selecting and Using Electric Motors. USDA Farmer Bulletin No. 2257, Washington, D.C.

9

Lighting

A good lighting system will place the proper amount and kind of light in the right places. The amount of light needed is dependent on the task being performed. The light needed to see while crossing a farmyard and that needed to operate a lathe are quite different. Having good lighting is vital to our health, comfort, convenience and safety.

The objective of this chapter will be to describe the basic concepts of lighting, types of light sources and methods for determining lighting needs such that decisions about lighting systems for farm buildings and farmstead areas can be made. Techniques described will be applicable to a wide variety of lighting problems. Lighting is a very broad and sometimes quite technical field, therefore this chapter will not be able to present the topic in great detail. For further information the reader is referred to the reference list at the end of the chapter, in particular the IES Lighting Handbook.

9.1 BASIC CONCEPTS OF LIGHT

Light is defined as visually evaluated radiant energy. Light is actually a small portion of the electromagnetic spectrum. Visible light varies in color, color being determined by the wavelength of the light. We are able to see objects because light is reflected from an object.

The amount of light received from a source per unit of power input is controlled by a number of variables. Some of the variables affect-

185

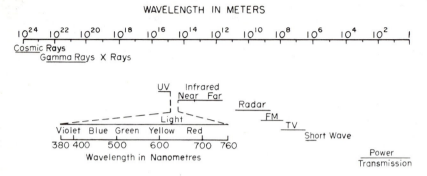

FIG. 9.1. ELECTROMAGNETIC SPECTRUM

ing lighting effectiveness include type of light sources (such as incandescent, fluorescent, or mercury vapor), how much the surroundings reflect or diffuse the light, and the distance from the lamp to the work.

Two interrelated measurements of illumination are important in consideration of light. The intensity of light at a point is measured in units of footcandles (fc) for the English system or in lux (lx) for the SI system. The quantity of light output by a source is measured in lumens (lm) for both systems.

If a candle is located in the center of a sphere having a radius of one foot, every point on the sphere is said to be illuminated to a level of one footcandle (fc) since each point is located one foot from the light source. Similarly, if the sphere has a radius of one meter, every

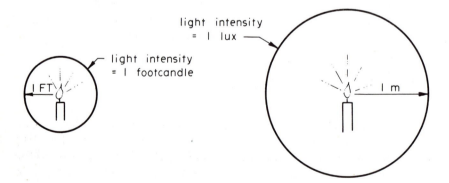

FIG. 9.2. DEFINING OF LIGHT INTENSITY

point on the sphere is said to be illuminated to a level of one lux (lx). It is important to note that both are measures of intensity of light at a point. No area units are involved. Conversion between footcandles and lux can be accomplished by the formula

$$1 \text{ fc} = 1 \text{ lux} \times 0.0929$$

or

$$1 \text{ lux} = 1 \text{ fc} \times 10.76.$$

To quantify the amount of light falling on a particular area the units of lumens (lm) are used. One lumen is the amount of light falling on each square foot of area illuminated with an intensity of one footcandle. To produce a level of illumination of 20 footcandles, 20 lumens for every square foot would be required or a level of 215 lux requires 215 lumens per square meter.

When designing lighting systems, it is important to understand the effect of distance on light. Light intensity varies as the square of the distance from the light source. To illustrate this characteristic of light, suppose a hole of 1 m² is cut in the sphere of 1 m radius shown earlier. With a light source of one candle in the center of the sphere, exactly 1 lumen of light will pass through the opening. If rounded screens are projected at distances of 2 m and 3 m from the source, as shown in Fig. 9.3, the illuminated area of the screens would be 4 m² and 9 m², respectively. Since the quantity of light remains constant (1 lumen), the second screen would receive 1/4 lm/m², the third would receive only 1/9 lm/m². It is clear that intensity is falling off as the square of the distance, $\dfrac{1}{2 \times 2} \dfrac{\ell m}{m^2}$ at 2 m and $\dfrac{1}{3 \times 3} \dfrac{\ell m}{m^2}$ at 3 m.

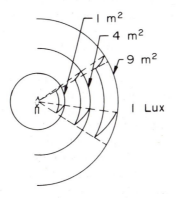

FIG. 9.3. LIGHT INTENSITY VARIES INVERSELY AS THE SQUARE OF DISTANCE FROM THE LIGHT SOURCE

9.2 TYPES OF LIGHT SOURCES

Types of light sources, including incandescent, fluorescent, and mercury and sodium vapor lights, will be discussed in the following sections. Characteristics of common sizes of each are given.

9.2.1 Incandescent Lamps

Incandescent means a glowing due to heat. Development of the incandescent lamp marked the beginning of efficient electric lighting. Table 9.1 traces the development of incandescent lights and the improvements in efficiency (lumens per watt) which have been obtained.

TABLE 9.1.   HISTORY OF INCANDESCENT LAMPS

1879 — Edison's first commercial lamp — carbon filament, 1.4 lumens per watt
1893 — Carbonized cellulose filament — 3.3 lumens per watt
1905 — Metallized carbon — 4 lumens per watt
1906 — Osmium and tantalum — 4.8 lumens per watt
1907 — Pressed or squired tungsten — 7.9 lumens per watt
1911 — Mazda drawn-tungsten wire, type B — 10 lumens per watt
1915 — 500, 300, 200, 100 watt sizes — 12.6 lumens per watt
Today — up to 26 lumens per watt

Figure 9.4 shows the construction of a typical incandescent bulb. Tungsten wire is generally used as the filament. When an electric current flows through the tungsten wire, the wire becomes warm due to the I^2R heating effect. If the resistance of the wire is high enough, the heating causes the wire to glow. This is the principle of an incandescent lamp. Most incandescent lamps are either evacuated or have an inert gas instead of a vacuum inside the bulb. The bulb of most lamps is coated or frosted to diffuse the light making it more appealing and reducing glare. Only about 6 to 12% of the energy emitted from an incandescent bulb will be in the visible range. Most of the radiation is in the infrared region. Most incandescent lamps used today have efficiencies of about 20 lumens per watt with larger bulbs being more efficient than smaller ones.

Typical initial output in lumens, lamp lumen depreciation factor (LLD) and average life expectancy ratings of common sizes of incandescent lamps are shown in Table 9.2. As the lamp ages, its inner surface becomes blackened from minute tungsten deposits and the lumen output drops somewhat. The lamp lumen depreciation given

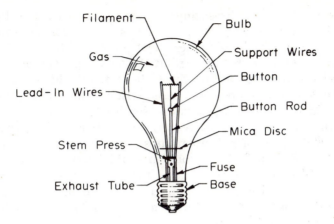

FIG. 9.4. CONSTRUCTION OF AN INCANDESCENT BULB

times the initial lumens output gives the output at 70% of rated life, i.e.

$$\begin{pmatrix}\text{Lamp Lumen}\\\text{Depreciation}\\\text{Factor}\end{pmatrix} \times \begin{pmatrix}\text{Initial}\\\text{Lumens}\\\text{Output}\end{pmatrix} = \begin{pmatrix}\text{Output at}\\70\%\text{ of}\\\text{Rated Life}\end{pmatrix}$$

TABLE 9.2. CHARACTERISTICS OF INCANDESCENT LAMPS — 115, 120, and 125 VOLT CIRCUITS.[1]

Rating in Watts	Initial Lumens Output	Lamp Lumen Depreciation[2] (%)	Average Life (hours)
25	230	79	2500
40	455	87.5	1500
60	860	93	1000
100	1740	90.5	750
150	2880	89	750
200	4000	89.5	750
300	6360	87.5	750
500	10,600	89	1000
750	17,000	89	1000

[1] Source IES Handbook (1972).
[2] Percent of initial light output at 70% of rated life.

9.2.2 Fluorescent Lamps

The development of fluorescent lamps, which started in 1938, began a new era in lighting. In many applications fluorescent lamps have several advantages over incandescent lamps including 1) high efficiency — two to three times as many lumens per watt which means they can operate at 50% or less of the cost of equivalent tungsten lighting systems, 2) less heat given off, 3) light produces less glare since the source is distributed over a larger area and 4) under ordinary operating conditions, five to ten times the life expectancy.

Fluorescent lamps do have characteristics which may limit their use. These include 1) temperature sensitivity, lamps intended for residential use operate properly only above 50° F, special lamps must be designed for cooler temperatures; 2) protective enclosures are needed in high humidity environments; 3) life expectancy is lowered by increasing the number of times the light is turned on and off, published ratings generally assume the light to be turned on three hours each time it is turned on; 4) a power factor of less than unity; and 5) a higher initial cost.

Fluorescent lamps are available in three main types, classified by their starting and operating circuit: hot cathode, preheat-starting; hot cathode, instant start; and cold cathode. The lamp consists of a glass tube fitted with special sockets for each type. The inside of the tube is coated with chemicals called phosphors which give off visible light when exposed to ultraviolet light.

Before a fluorescent lamp will "light", it requires a momentary high volt, higher than the line voltage, in order to establish an arc discharge through the gas inside the tube. This is developed by adding an inductance coil (a ballast) in series with one side of the line. Figure 9.5 shows two possible circuits for ballast coils. When the switch is closed, current flows from the line through the ballast, the

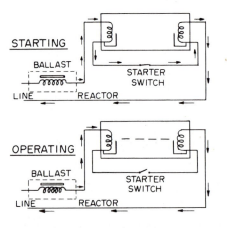

FIG. 9.5.   TWO BALLAST COIL CIRCUITS FOR FLUORESCENT LAMPS

first cathode, the starter, the second cathode and back to the line. The time for which the current heats the cathode is determined by the starter. When the starter opens, the inductive surge from the ballast strikes an arc between the cathodes. Heat from the arc evaporates mercury within the tube which creates a low resistance path between the cathodes through the mercury vapor. The ballast then performs the function of limiting the current flow through the tube.

As the electrons move at high rates of speed from one cathode to the other, they are likely to collide with mercury vapor atoms. During the collision electrons from the mercury vapor may be temporarily moved out of their shell locations. As they return to their locations they emit ultraviolet radiation. This ultraviolet radiation is transformed to visible light by the phosphor coating of the tube. (Fig. 9.6).

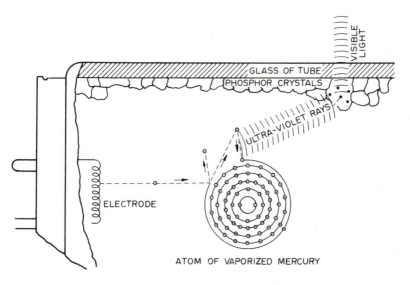

FIG. 9.6.   FLUORESCENT LAMP OPERATION

Like other types of lamps, fluorescent lamps deteriorate in light output as they age. The curve shown in Fig. 9.7 shows the typical range of fluorescent lamp output over time.

Typical values for initial lamp lumens, lamp lumen depreciation (LLD), and average life are given in Table 9.3 for common fluorescent lamp sizes. Different types of phosphors yield the three different types of lights shown and their corresponding differences in initial lumens.

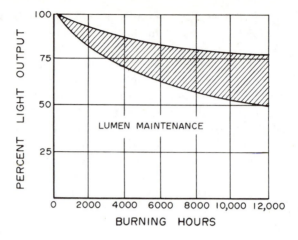

FIG. 9.7.   FLUORESCENT LIGHT OUTPUT VERSUS BURNING TIME

TABLE 9.3.   CHARACTERISTICS OF TYPICAL FLUORESCENT LAMPS (PREHEAT STARTING)

Nominal Watts	20	30	40	90
Nominal Lengths (inches)	24	36	48	60
Rated Life	7500–9000	7500	18,000	9000
Lamp Lumen Depreciation (LLD)[1]	85	79	82	85
Initial Lumens				
Cool White	1250	2190	3200	6350
Warm White	1270	2220	3250	6450
Deluxe Warm White	855	1500	2190	4350

Source: IES Lighting Handbook (1972).

[1] Percent of initial light output at 70% of rated life at 3 hours per start

9.2.3 Mercury and Sodium Vapor Lamps

Mercury lamps combine the relatively small size of incandescent lamps with the long life and high efficiency of fluorescent lamps. The high wattage-to-size ratio makes them particularly adaptable for industrial and street lighting, floodlighting, and other outdoor lighting. Mercury vapor lamps are electric discharge lamps. As such, they produce light by passing a current through a gas vapor under pressure rather than heating a filament as an incandescent lamp. Like fluorescents, mercury lamps require ballasts to regulate current flow. Mercury lamps produce a greenish-blue light with efficiencies of 30 to 65 lumens per watt.

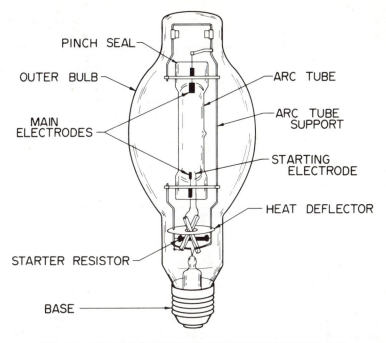

PINCH SEAL

OUTER BULB

MAIN
ELECTRODES

ARC TUBE

ARC TUBE
SUPPORT

STARTING
ELECTRODE

HEAT DEFLECTOR

STARTER RESISTOR

BASE

FIG. 9.8.   THE BASIC COMPONENTS OF THE MERCURY LAMP

Figure 9.8 illustrates the basic components of the mercury lamp. Most mercury lamps consist of an arc tube enclosed within a protective outer tube. The arc tube contains the essential components — electrodes, mercury vapor, and argon gas. The outer bulb helps maintain more constant temperature as well as protect the arc tube and other parts from the atmosphere.

Sodium vapor lamps are very similar to mercury lamps only using sodium vapor and neon gas. The low pressure sodium vapor lamp produces almost all its radiation at a wavelength of 5893 angstroms. This gives it its distinct golden-yellow color light during operation. This wavelength is near the peak sensitivity of the eye. As a result, low pressure sodium vapor lamps have a high efficiency (approximately 50 lumens/watt). They are used primarily for highway and street lighting. High pressure sodium lamps produce light of a golden-white color with all visible frequencies present. These lamps have efficiencies of about 110 lumens per watt. Because of their very high efficiency, high pressure sodium lights are finding wide use for applications such as factory and warehouse lighting.

Characteristics of typical mercury and sodium lamps are given in Table 9.4.

194   FUNDAMENTALS OF ELECTRICITY FOR AGRICULTURE

TABLE 9.4. CHARACTERISTICS OF TYPICAL MERCURY AND SODIUM VAPOR LAMPS

	Lamp Watts	Initial Lumens	Lamp Lumen Depreciation (%)	Rated Life (hours)
Mercury Lamps	40	1,165	—	16,000
	75	2,700	83	10,000
	100	3,700	83	10,000
	175	7,535	79	24,000+
	250	11,270	80	24,000+
	400	20,750	76	24,000+
High Pressure Sodium	250	25,500	88	10,000
	400	47,000	86	15,000
Low Pressure Sodium	60	6,000	—	15,000
	200	25,000	—	15,000

Source: IES Lighting Handbook (1972).

9.3 LIGHTING REQUIREMENTS

As complete a knowledge as possible of the visual requirements and physical environment for an area is necessary to correctly specify the illumination level and equipment needed. One important factor is the level of illumination required for the area. Recommended levels of illumination for different areas and tasks are given to assist designers in references such as the IES Lighting Handbook, American National Standards Institute, and the Agricultural Engineers' Yearbook. Tables 9.5 through 9.8 give some of the recommendations for levels of illumination in farm and home situations.

9.4 TYPES OF LIGHTING SYSTEMS

Lighting systems are often classified in two manners. First, in accordance with their layout or location with respect to the task area, they are classed as general lighting, localized lighting, and local (supplementary) lighting. They are also classified in accordance with the type of luminaries[1] used as direct, semi-direct, general diffuse (direct-indirect), semi-indirect, and indirect.

---

[1] A luminaire is a complete lighting unit and is made of a light source together with other integral parts such as a globe, reflector, socket, housing, etc.

9.4.1 Classification by Layout and Location

Lighting systems which provide an approximately uniform level of illumination over the entire area are called general lighting systems. The chief advantage of general lighting is that it permits flexibility in task location. The luminaires are usually arranged in a symmetric plan to fit the physical characteristics of the area.

A localized general lighting system has the luminaires located such that lighting is more concentrated at designated task areas. Caution must be exercised so that good general lighting without shadows or glare is maintained.

A local lighting system provides lighting only over a relatively small area occupied by the task and its immediate surroundings. This is an economical means of providing higher illumination levels over a small area, and it usually permits some adjustment of lighting to suit the requirements of the individual. Table lamps and spot lights are exam-

TABLE 9.5. RECOMMENDED ILLUMINATION FOR POULTRY FARM INDUSTRY TASKS

Areas and Visual Tasks	Minimum Illumination At Any Time Footcandles	Lux
Brooding, Production, and Laying Houses		
Feeding, inspection, and cleaning	20	220
Charts and records	30	320
Thermometers, thermostats, and time clock	50	540
Hatcheries		
General area and loading platform	20	220
Inside incubators	30	320
Dubbing station	150	1600
Sexing	1000	10760
Egg Handling, Packing, and Shipping		
General cleanliness	50	540
Egg quality inspection	50	540
Loading platform, egg storage area, etc.	20	220
Egg Processing		
General lighting	70	750
Fowl Processing Plant		
General (excluding killing and unloading area)	70	750
Government inspection station and grading stations	100	1080
Unloading and killing area	20	220
Feed Storage		
Grain, feed rations	10	110
Processing	10	110
Charts and records	30	320

Source: Agricultural Engineers' Yearbook (1977).

TABLE 9.6.   RECOMMENDED ILLUMINATION FOR DAIRY FARMS

	Minimum Illumination At Any Time	
Areas and Visual Tasks	Footcandles	Lux
Milking Operation Area (milking parlor & stall barn)		
General	20	220
Cow's udder	50	540
Milk Handling Equipment and Storage Area (milk house or milk room)		
General	20	220
Washing area	100	1080
Bulk tank interior	100	1080
Loading platform	20	220
Feeding Area (stall barn feed alley, pens, and loose housing feed area)	20	220
Feed Storage Area, Forage		
Haymow	3	30
Hay inspection area	20	220
Ladders and stairs	20	220
Silo	3	30
Silo room	20	220
Feed Storage Area, Grain and Concentrate		
Grain bin	3	30
Concentrate storage area	10	110
Feed Processing Area	10	110
Livestock Housing Area (community, maternity, individual calf pens, and loose-housing holding and resting areas)	7	80

Source: Argicultural Engineers' Yearbook (1977).

TABLE 9.7.   RECOMMENDED ILLUMINATION FOR GENERAL AREAS ASSOCIATED WITH DAIRY AND POULTRY FACILITIES

	Minimum Illumination At Any Time	
Areas and Visual Tasks	Footcandles	Lux
Machine Storage		
Garage and machine shed	5	50
Farm Shop		
Active storage area	10	110
General shop	30	320
Rough bench machine work	50	540
Medium bench machine	100	1080
Miscellaneous		
Farm office	70	750
Restrooms	30	320
Pumphouse	20	220
Exterior		
General inactive areas	0.2	2
General active areas (paths, rough storage, barn lots)	1	10
Service area (fuel storage, shop, feed lots, building entrances)	3	30

Source: Agricultural Engineers' Handbook (1977).

TABLE 9.8.   RECOMMENDED LEVELS OF ILLUMINATION FOR RESIDENCES

	Footcandles	Lux
General Lighting		
Conversation, relaxation, entertainment	10	110
Passage areas for safety	10	110
Areas involving visual tasks, other	30	320
than kitchen		
Kitchen	50	540
Specific Visual Tasks		
Dining	15	160
Grooming, shaving, make-up	50	540
Ironing	50	540
Kitchen duties		
Food preparation and cleaning	150	1600
Serving and other non-critical tasks	50	540
Laundry		
Preparation, sorting, inspection	50	540
Tub area-soaking, tinting	50	540
Washer and dryer areas	30	320
Sewing (hand and machine)		
Dark fabrics	200	2200
Medium fabrics	100	1100
Light fabrics	30	320
Study	70	750
Table games	30	320

Source: IES Lighting Handbook (1972).

ples of local lighting. Local lighting by itself is seldom desirable. General lighting should supply at least 20% of the lighting such that lighting blends across the area and bright spots are not created.

9.4.2 Classification by Luminaire Type

Luminaires for general lighting are classified by the percentage of total light output directed above and below horizontal. Categories are summarized in Fig. 9.9. Direct systems generally have the highest utilization factor and are most widely used for general purpose lighting. Indirect systems tend to give better glare control and are widely used for residential and commercial lighting.

9.5 LIGHTING CALCULATIONS FOR INTERIOR AREAS

The method for general lighting calculations presented here is called the Zonal Cavity Method. This method gets its name from the procedure of dividing the room into zones (or cavities). This method is recommended by the Illuminating Engineers' Society because of its flexibility and application to a wide range of interior space.

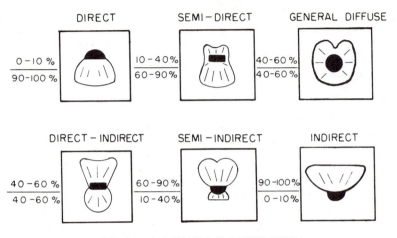

FIG. 9.9.   LUMINAIRE CLASSIFICATION

The function of the lighting calculations are generally to determine how many luminaires are needed to provide average illumination throughout the room.

Theoretically, the level of illumination could be calculated by dividing the source output by the area to be illuminated, i.e.,

$$\text{Illumination} = \frac{\text{Lumens Produced}}{\text{Area}}.$$

If the area is in square meters, the illumination is in units of lux (lumens per square meter). If the area is in square feet, the illumination is in units of foot-candles (lumens per square foot).

In actual application, a portion of the light produced will be lost before it reaches the work plane. Losses occur both in the luminaire itself and at the room surfaces. A coefficient of utilization, CU, which represents the portion of light which reaches the work area can be included in the formula to yield

$$\text{Initial Illumination} = \frac{\text{Initial Lamp Lumens} \times \text{Coefficient of Utilization}}{\text{Area}}$$

Since our objective is to design a system which will maintain a certain minimum level of illumination over time, an additional factor termed Light Loss Factor, LLF, must be included. Estimated deterioration in the light source and estimated losses from dirt collection on the luminaires and room surfaces are the major factors in determining the Light Loss Factor, LLF. The illumination formula now becomes,

$$\text{Maintained Illumination} = \frac{\text{ILL} \times \text{CU} \times \text{LLF}}{\text{Area}}$$

where
- ILL = Initial Lamp Lumens
- CU = Coefficient of Utilization
- LLF = Light Loss Factor

If the desired illumination level and luminaire to be used are known, the formula can be rearranged to yield the area per luminaire (and hence the spacing between luminaires).

$$\frac{\text{Area per}}{\text{Luminaire}} = \frac{\text{Lamp Lumens per Luminaire} \times \text{CU} \times \text{LLF}}{\text{Maintained Illumination}}$$

For an established lighting system, the maintained illumination can be calculated by,

$$\frac{\text{Maintained}}{\text{Illumination}} = \frac{\text{Lamp Lumens per Luminaire} \times \text{CU} \times \text{LLF}}{\text{Area per Luminaire}}$$

The following two sections deal with determining the CU and LLF factors.

9.5.1 Determining Coefficient of Utilization

For determining the Coefficient of Utilization, the first step is to establish the cavity ratios and the effective cavity reflectances. The effect of room proportions, luminaire suspension length, and work-plane height upon the coefficient of utilization are accounted for by the Room Cavity Ratio, RCR, the Ceiling Cavity Ratio, CCR, and the Floor Cavity Ratio, FCR. These ratios are determined by dividing the room into three cavities as shown in Fig. 9.10. Then, using the room dimensions and each cavity depth, each cavity ratio can be found from the formula:

$$\text{Cavity Ratio} = \frac{5h\,(\text{Room Length} + \text{Room Width})}{\text{Room Length} \times \text{Room Width}}$$

where
- $h = h_{RC}$ for the Room Cavity Ratio, RCR
- $= h_{CC}$ for the Ceiling Cavity Ratio, CCR
- $= h_{FC}$ for the Floor Cavity Ratio, FCR

The Coefficient of Utilization factor also takes into account the reflectance of the floor, walls and ceiling. Table 9.9 gives some

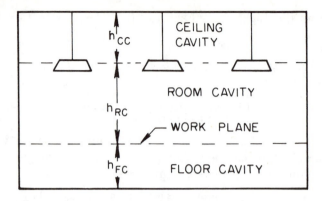

FIG. 9.10. THE THREE CAVITIES IN THE ZONAL CAVITY
METHOD

recommended reflectance values for use in determining lighting for farm buildings.

TABLE 9.9. RECOMMENDED MATTE REFLECTANCE VALUES FOR LIGHTING FARM BUILDINGS

Surface	Reflectance, %
Ceiling	80 to 90
Walls	40 to 60
Desk and bench tops, machines and equipment	25 to 45
Floors	20 minimum

Source: ASAE, EP 344 (1971).

Table 9.10 provides a means for converting the combination of initial wall and ceiling, and wall and floor reflectances into a single Effective Ceiling Cavity Reflectance, ρ_{CC} and a single Effective Floor Cavity Reflectance, ρ_{FC}. Note that for surface mounted and recessed luminaires (CCR = 0), the ceiling reflectance may be used for ρ_{CC}.

Coefficient of Utilization for selected luminaire types can now be determined using Table 9.11. Table 9.11 is a tabulation of Coefficients of Utilization for representative luminaire types to be used in Zonal Cavity Method calculations. The coefficients are for an Effective Floor Cavity Reflectance, ρ_{FC}, of 20%. However, the Coefficient of Utilization obtained from Table 9.11 can be adjusted to other floor reflectances (ρ_{FC}) if necessary by multiplying factors found in Table 9.12.

Figure 9.11 contains a Coefficient of Utilization worksheet to assist in the process of determing CU. Its use will be demonstrated

Room Identification _____

Step 1: Fill in sketch.

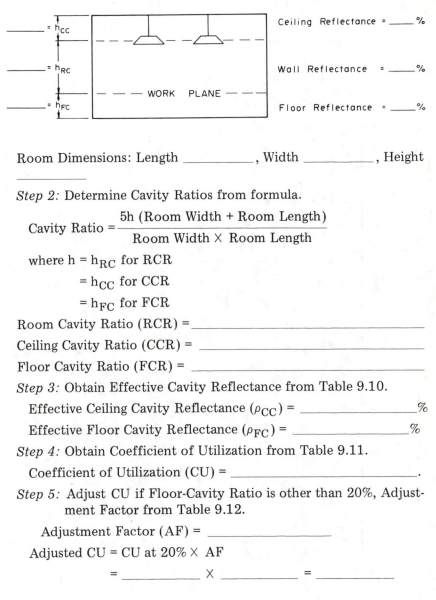

Room Dimensions: Length _____, Width _____, Height

_____

Step 2: Determine Cavity Ratios from formula.

$$\text{Cavity Ratio} = \frac{5h \,(\text{Room Width} + \text{Room Length})}{\text{Room Width} \times \text{Room Length}}$$

where $h = h_{RC}$ for RCR

$\quad\quad = h_{CC}$ for CCR

$\quad\quad = h_{FC}$ for FCR

Room Cavity Ratio (RCR) = _____

Ceiling Cavity Ratio (CCR) = _____

Floor Cavity Ratio (FCR) = _____

Step 3: Obtain Effective Cavity Reflectance from Table 9.10.

Effective Ceiling Cavity Reflectance (ρ_{CC}) = _____%

Effective Floor Cavity Reflectance (ρ_{FC}) = _____%

Step 4: Obtain Coefficient of Utilization from Table 9.11.

Coefficient of Utilization (CU) = _____.

Step 5: Adjust CU if Floor-Cavity Ratio is other than 20%, Adjustment Factor from Table 9.12.

Adjustment Factor (AF) = _____

Adjusted CU = CU at 20% × AF

$\quad\quad\quad = $ _____ × _____ = _____

FIG. 9.11. COEFFICIENT OF UTILIZATION WORKSHEET

TABLE 9.10.   PERCENT EFFECTIVE CEILING OR FLOOR CAVITY REFLECTANCES FOR VARIOUS REFLECTANCE COMBINATIONS

Per Cent Base[1] Reflectance	90										80										70										60										50									
Per Cent Wall Reflectance	90	80	70	60	50	40	30	20	10	0	90	80	70	60	50	40	30	20	10	0	90	80	70	60	50	40	30	20	10	0	90	80	70	60	50	40	30	20	10	0	90	80	70	60	50	40	30	20	10	0
Cavity Ratio																																																		
0.2	89	88	88	87	86	85	85	84	84	82	79	78	78	77	77	76	75	75	74	72	70	69	68	68	67	67	66	66	65	64	60	59	59	58	57	56	56	55	55	53	50	50	49	49	48	48	47	46	46	44
0.4	88	87	86	85	84	83	81	80	79	76	79	77	76	75	73	71	70	68	66	63	69	68	67	65	64	62	61	59	57	54	60	59	57	56	55	53	51	51	50	46	50	49	48	47	46	45	44	43	42
0.6	87	86	84	82	80	79	77	76	74	73	78	76	75	73	71	70	68	66	65	63	69	67	65	64	63	61	59	58	57	54	60	58	57	55	53	51	51	50	43	50	48	47	45	44	43	42	41	38
0.8	87	85	82	80	77	75	73	71	69	67	77	75	73	71	69	67	65	63	60	57	68	67	65	63	61	59	56	53	52	50	58	57	56	55	53	51	48	47	46	43	50	48	47	45	44	43	42	38	36
1.0	86	83	80	77	75	73	72	69	66	64	62	77	74	72	69	67	65	62	60	57	55	68	65	62	60	58	55	53	52	50	47	57	56	55	53	51	48	45	44	43	41	48	46	44	43	41	38	37	36	34
1.2	85	82	78	75	72	69	66	63	60	57	76	73	70	67	64	61	59	57	54	51	67	64	61	59	57	54	50	48	46	44	59	56	54	51	49	46	44	41	39	36	39	41	43	41	39	36	35	34	29
1.4	85	80	77	73	69	65	62	59	57	52	76	72	68	65	62	58	55	53	50	48	67	63	60	58	55	51	47	45	43	41	59	56	53	49	47	44	41	39	38	35	50	47	45	42	40	38	35	34	27
1.6	84	79	75	71	67	63	60	56	53	50	75	71	67	63	60	57	53	50	47	44	67	62	59	56	53	47	45	43	41	38	59	55	52	48	45	42	39	37	36	33	50	47	44	41	39	36	33	32	26
1.8	83	78	73	69	64	60	56	53	50	48	74	69	64	60	56	52	49	47	45	41	66	61	58	54	51	46	42	40	38	35	58	55	51	47	44	40	37	35	33	31	50	46	43	40	38	35	33	28	24
2.0	81	77	72	67	62	56	53	50	47	43	74	69	64	60	56	52	48	45	40	37	66	60	56	52	49	45	42	40	38	33	58	54	50	46	43	39	36	34	32	29	50	46	42	40	37	34	30	28	24
2.2	82	76	70	65	59	54	50	47	44	40	76	70	65	59	54	49	45	42	38	35	70	60	55	51	48	43	38	36	34	32	53	42	37	34	31	29	27	25	23	22	50	46	42	38	33	31	27	25	23	22
2.4	82	75	69	64	58	53	48	45	41	37	75	69	64	58	53	47	43	40	36	33	69	60	54	50	46	41	37	35	32	30	53	42	36	32	30	28	26	23	21	20	50	46	42	37	33	30	26	23	21
2.6	81	74	67	62	56	51	46	42	38	35	74	67	60	55	50	45	41	38	34	31	64	59	54	49	45	40	35	33	30	26	53	43	35	31	29	26	24	22	20	16	50	46	41	37	34	30	26	23	21	20
2.8	81	73	66	60	54	49	44	40	36	34	73	66	59	53	47	42	38	34	32	29	59	53	49	44	42	38	33	30	28	21	58	47	34	32	28	24	22	21	19	15	50	45	41	36	33	29	25	22	19	17
3.0	80	72	64	58	52	47	42	38	34	30	72	63	58	52	46	42	36	32	29	25	58	54	49	45	42	37	32	29	27	17	58	46	40	36	32	28	25	20	18	14	50	45	40	36	32	28	24	21	19	17
3.2	79	71	63	56	50	45	40	36	32	28	72	65	57	51	45	40	35	33	28	25	62	55	51	46	40	36	31	28	25	23	57	45	40	36	31	27	23	22	18	52	50	44	39	35	31	27	23	20	18	16
3.4	79	70	62	54	48	43	38	34	31	28	71	64	56	50	44	39	34	32	28	24	61	54	46	44	41	35	30	26	23	21	57	45	39	35	30	26	23	20	17	50	44	38	35	31	27	23	19	16	15
3.6	79	69	61	53	47	42	36	33	29	25	71	63	54	48	43	38	32	30	25	23	60	54	44	43	40	33	28	24	21	20	56	44	39	34	29	25	21	18	16	50	44	38	34	30	26	23	18	16	14
3.8	78	69	60	51	45	40	35	31	27	23	70	62	53	47	42	36	31	28	24	21	58	53	43	41	38	32	27	24	20	18	57	43	38	33	28	24	21	17	15	50	43	38	33	29	25	22	17	15	13
4.0	77	68	58	51	44	38	33	29	25	22	70	61	53	46	40	35	30	26	23	20	58	52	42	40	37	31	26	23	19	17	57	42	37	32	28	24	19	17	15	12	50	42	35	30	25	21	17	15	12
4.2	77	62	57	50	43	37	32	28	24	21	69	60	52	44	39	34	29	25	22	18	55	47	41	35	30	25	22	19	16	56	49	42	37	32	27	23	20	17	14	12	50	43	37	32	28	24	20	17	14	12
4.4	76	61	49	42	36	31	27	23	20	69	60	51	44	38	32	27	23	20	17	54	46	40	34	29	24	21	18	16	56	49	42	36	31	27	22	19	16	13	11	50	43	37	32	27	23	19	16	13	11
4.6	76	60	47	40	35	30	26	22	19	69	59	50	43	37	32	26	23	19	15	54	45	39	33	28	24	20	17	14	56	49	41	35	30	26	21	18	15	12	09	50	43	36	31	26	22	18	15	12	09
4.8	75	59	46	39	33	28	24	21	18	68	58	49	42	36	31	25	22	18	14	53	45	38	31	27	23	19	16	12	55	48	41	34	29	25	20	17	15	12	07	50	42	35	30	25	21	17	15	12	08
5.0	75	59	45	38	33	28	24	20	16	68	58	49	40	35	30	25	21	18	14	52	43	36	30	26	22	19	16	12	55	48	40	34	28	24	20	17	14	11	07	50	42	35	29	24	20	17	14	12	08
6.0	73	61	49	41	34	29	24	20	16	11	66	55	44	38	31	27	22	19	15	10	51	41	35	28	24	19	16	13	09	51	45	37	31	25	21	17	14	11	07	50	42	34	29	23	19	15	13	10	06
7.0	70	58	45	38	30	27	21	18	14	08	64	53	42	36	28	23	19	16	11	06	48	38	32	26	22	17	14	11	06	48	43	35	30	24	19	16	13	10	05	49	41	32	27	21	18	14	11	08	05
8.0	68	55	42	35	27	23	18	15	12	06	62	50	38	32	25	21	17	13	10	05	45	35	29	23	19	15	12	10	04	45	42	33	28	22	18	15	12	10	05	49	40	30	25	19	16	12	10	07	03
9.0	66	52	38	31	25	21	16	14	11	05	61	49	36	30	23	19	15	13	11	04	43	33	27	21	18	14	12	09	04	40	31	26	20	16	14	11	09	07	03	48	39	29	24	18	15	11	09	07	03
10.0	65	51	36	29	22	19	15	11	09	04	59	46	33	27	21	18	14	11	08	03	42	31	25	19	16	12	10	08	57	45	33	27	21	18	14	12	09	07	02	47	37	27	22	17	14	10	08	06	02

[1] Ceiling, floor, or floor of cavity.

TABLE 9.10. (CONTINUED)

Per Cent Base¹ Reflectance	40										30										20										10										0									
Per Cent Wall Reflectance	90	80	70	60	50	40	30	20	10	0	90	80	70	60	50	40	30	20	10	0	90	80	70	60	50	40	30	20	10	0	90	80	70	60	50	40	30	20	10	0	90	80	70	60	50	40	30	20	10	0
Cavity Ratio																																																		
0.2	40	40	39	39	38	38	37	36	36		31	30	30	30	29	29	29	28	28	27	21	20	20	20	20	20	19	19	19	17	11	11	11	10	10	10	10	09	09	09	02	02	01	01	01	01	00	00	0	0
0.4	41	40	39	38	37	36	35	34	34		31	31	30	29	28	27	26	25			22	21	20	20	19	19	18	18	16		12	11	11	11	11	10	10	09	09	08	04	03	03	02	02	02	01	01	00	0
0.6	41	40	39	38	37	36	34	33	32	31	32	31	30	29	28	27	26	25	25	23	23	21	21	20	19	18	17	16	15		13	13	12	11	11	10	10	09	08	08	05	05	04	03	03	02	02	01	01	0
0.8	42	40	38	37	36	35	33	32	31		32	31	30	29	28	27	25	24	23	22	24	22	21	20	19	18	17	16	15		13	13	12	12	11	11	10	09	08	07	07	06	05	04	04	03	02	02	01	0
1.0	42	40	38	37	35	33	32	31	29	27	32	31	30	29	27	25	24	22	21	20	25	23	22	20	18	17	16	15	13		15	14	13	12	12	11	10	09	08	07	08	07	06	05	04	03	02	02	01	0
1.2	42	40	38	36	34	32	30	29	27	25	33	32	30	28	27	25	23	22	21	19	25	24	22	21	19	17	16	14	12		17	15	14	13	12	11	10	09	07	06	10	08	07	06	05	05	03	02	01	0
1.4	42	39	37	35	33	31	29	27	25	23	33	32	31	29	27	25	23	21	20	18	26	24	22	21	18	17	15	13	11		18	16	14	13	12	11	10	09	07	06	11	09	08	07	06	04	03	02	01	0
1.6	42	39	37	35	32	30	28	26	24	22	34	33	31	29	27	25	23	21	18	17	26	24	23	21	18	17	14	12	10		19	17	15	14	13	11	09	08	07	06	12	10	09	07	06	05	04	03	02	0
1.8	42	39	36	34	31	29	27	25	23	21	35	33	31	29	27	25	23	20	18	16	27	25	23	22	18	15	14	12	10		19	17	16	15	13	11	09	08	06	05	13	11	09	08	07	05	04	03	02	0
2.0	42	39	36	34	31	28	26	24	22	19	35	33	31	29	27	25	22	20	18	14	28	25	23	23	18	15	13	11	09		20	18	16	15	13	12	11	09	08	07	14	12	10	09	07	05	04	03	02	0
2.2	42	39	36	33	30	27	24	22	19	18	36	35	33	31	29	26	24	22	19	17	28	25	23	23	16	14	12	10	09		21	19	17	16	13	11	09	07	06	05	15	13	11	09	07	06	05	03	02	0
2.4	42	39	35	32	29	27	24	21	18	17	36	35	33	31	29	26	24	22	18	16	29	26	24	23	16	14	11	10	08		21	19	17	16	14	12	10	08	06	05	16	13	11	09	08	06	05	04	02	0
2.6	43	39	35	32	29	26	23	20	17	15	36	35	32	30	28	26	23	21	18	15	29	26	24	23	16	14	11	09	07		22	20	18	16	14	12	09	07	06	04	17	14	12	10	08	06	05	04	02	0
2.8	43	39	35	32	28	25	22	19	16	14	37	35	32	30	28	25	23	21	17	15	30	27	25	23	15	13	11	09	07		23	20	18	17	14	11	09	06	05	03	17	15	12	10	08	07	05	04	02	0
3.0	43	39	35	31	27	24	21	18	15	13	37	33	33	29	25	22	20	18	16	12	30	27	25	23	17	13	11	09	07		24	21	18	17	13	11	09	06	05	03	18	16	13	11	09	08	05	03	02	0
3.2	43	39	35	31	27	23	20	17	15	13	38	33	33	29	25	22	20	18	16	12	31	27	23	23	15	12	11	09	06		25	21	18	16	13	11	09	06	05	03	19	16	14	11	09	07	05	03	02	0
3.4	43	39	34	30	26	23	20	17	14	11	39	33	33	29	24	21	20	17	14	09	31	27	23	23	15	12	10	08	06		26	22	18	16	12	09	09	06	04	04	20	17	14	12	09	07	04	03	02	0
3.6	44	39	34	30	26	23	19	16	14	11	39	33	32	29	24	21	19	16	13	09	32	28	23	23	15	11	10	08	05		26	22	20	17	14	11	09	06	04	03	20	17	15	12	10	08	06	04	02	0
3.8	44	38	33	29	25	22	19	16	13	10	39	33	33	28	24	21	18	15	13	10	32	28	24	23	15	11	10	08	05		27	23	19	17	13	10	08	05	04	02	21	18	15	12	10	08	06	04	02	0
4.0	44	38	33	29	25	21	18	15	12	10	39	33	33	28	24	18	18	14	12	07	33	28	24	23	14	11	09	07	05		27	23	20	17	14	11	09	06	04	02	22	18	15	13	10	08	06	04	02	0
4.2	44	38	33	29	24	21	17	15	12	10	38	33	29	24	20	17	14	12	09	07	33	28	24	23	14	11	09	07	04		28	24	20	17	14	11	09	06	04	02	22	19	16	13	11	08	06	04	02	0
4.4	44	38	33	28	24	20	17	14	11	09	38	33	29	24	20	17	13	11	08	06	34	29	24	23	14	11	09	07	04		28	24	20	17	14	11	08	06	04	02	23	19	16	13	11	08	06	04	02	0
4.6	44	38	32	28	23	19	16	13	10	08	39	33	28	24	19	16	13	10	08	06	35	29	24	23	13	10	08	06	04		29	25	20	17	13	10	08	06	04	02	23	20	17	13	11	08	06	04	02	0
4.8	44	38	32	27	22	19	15	13	10	08	39	33	28	24	19	16	13	10	08	05	35	29	24	23	13	10	08	06	04		29	25	21	18	13	10	08	06	04	02	24	20	17	14	11	08	06	04	02	0
5.0	45	38	31	27	22	19	15	12	10	07	39	33	28	24	19	15	12	10	08	05	35	29	24	23	11	09	08	06	04		30	25	20	17	12	10	07	06	04	02	25	21	17	14	11	08	06	04	02	0
6.0	44	37	30	25	20	17	13	11	08	05	39	33	27	23	18	15	11	09	06	04	36	30	24	23	13	10	08	05	02		31	26	21	18	14	09	07	05	03	01	27	23	18	15	12	09	06	04	02	0
7.0	44	36	29	24	19	16	12	10	07	04	40	33	26	22	17	10	10	08	05	03	36	30	23	23	12	09	07	04	01		32	27	21	17	13	09	07	05	03	01	28	24	19	15	12	09	06	04	02	0
8.0	44	35	28	23	18	15	11	08	06	03	40	33	26	21	16	09	07	04	02	01	37	30	23	23	11	08	06	03	01		33	27	21	17	13	10	07	05	03	01	30	25	20	15	13	11	08	06	04	0
9.0	44	35	27	22	18	13	10	08	05	02	40	33	25	20	15	12	09	05	01		37	30	23	23	14	11	08	05	01		34	28	21	17	13	10	07	05	02	01	31	25	20	15	12	09	06	04	02	0
10.0	43	34	25	20	15	12	08	07	05	02	40	33	24	19	14	11	08	06	03	01	37	29	22	18	13	10	07	05	02		34	28	21	17	12	10	07	05	02	01	31	25	20	15	12	09	06	04	02	0

Source: IES Lighting Handbook (1972).
¹ Ceiling, floor, or floor of cavity.

TABLE 9.11. COEFFICIENT OF UTILIZATION GUIDE

Typical Luminaire	Maint. Cat.	Maximum S/MH Guide	RCR	ρcc → 80, ρw 50	30	10	70, 50	30	10	50, 50	30	10	30, 50	30	10	10, 50	30	10	0
Porcelain-enameled ventilated standard dome with incandescent lamp (0%↑, 83½%↓)	IV	1.3	0	.99	.99	.99	.97	.97	.97	.92	.92	.92	.88	.88	.88	.85	.85	.85	.83
			1	.88	.85	.82	.86	.83	.81	.83	.80	.78	.79	.78	.76	.77	.75	.73	.72
			2	.78	.73	.68	.76	.72	.67	.73	.69	.66	.71	.67	.64	.68	.65	.63	.61
			3	.69	.62	.57	.67	.61	.57	.65	.60	.56	.63	.58	.55	.61	.57	.54	.52
			4	.61	.54	.49	.60	.53	.48	.58	.52	.48	.56	.51	.47	.54	.50	.46	.45
			5	.54	.47	.41	.53	.46	.41	.51	.45	.41	.50	.44	.40	.48	.43	.40	.38
			6	.48	.41	.35	.47	.40	.35	.46	.39	.35	.44	.39	.34	.43	.38	.34	.32
			7	.43	.35	.30	.42	.35	.30	.41	.34	.30	.39	.34	.30	.38	.33	.29	.28
			8	.38	.31	.26	.38	.31	.26	.37	.30	.26	.36	.30	.26	.35	.30	.26	.24
			9	.35	.28	.23	.34	.27	.23	.33	.27	.23	.32	.27	.23	.31	.26	.22	.21
			10	.31	.25	.20	.31	.24	.20	.30	.24	.20	.29	.24	.20	.29	.23	.20	.18
Wide distribution unit with lens plate and inside frost lamp (0%↑, 53½%↓)	V	1.4	0	.63	.63	.63	.62	.62	.62	.59	.59	.59	.56	.56	.56	.54	.54	.54	.53
			1	.58	.56	.54	.57	.55	.54	.54	.53	.52	.52	.51	.50	.50	.50	.49	.48
			2	.53	.50	.48	.52	.49	.47	.50	.48	.46	.48	.47	.45	.47	.45	.44	.43
			3	.48	.45	.42	.47	.44	.42	.46	.43	.41	.44	.42	.40	.43	.41	.40	.39
			4	.44	.40	.37	.43	.40	.37	.42	.39	.37	.41	.38	.36	.40	.38	.36	.35
			5	.40	.36	.33	.39	.36	.33	.38	.35	.33	.37	.35	.32	.36	.34	.32	.31
			6	.36	.32	.30	.36	.32	.29	.35	.32	.29	.34	.31	.29	.33	.31	.29	.28
			7	.33	.29	.26	.33	.29	.26	.32	.28	.26	.31	.28	.26	.30	.28	.26	.25
			8	.30	.26	.23	.30	.26	.23	.29	.26	.23	.28	.25	.23	.28	.25	.23	.22
			9	.27	.23	.21	.27	.23	.21	.26	.23	.21	.26	.23	.20	.25	.22	.20	.19
			10	.25	.21	.18	.25	.21	.18	.24	.21	.18	.24	.20	.18	.23	.20	.18	.17
Recessed unit with dropped diffusing glass (1½%↑, 50½%↓)	V	1.3	0	.61	.61	.61	.60	.60	.60	.57	.57	.57	.54	.54	.54	.51	.51	.51	.50
			1	.53	.51	.48	.52	.50	.47	.49	.47	.46	.47	.45	.44	.45	.44	.42	.41
			2	.46	.42	.39	.45	.42	.39	.43	.40	.38	.41	.39	.37	.39	.37	.35	.34
			3	.40	.36	.33	.40	.35	.32	.38	.34	.31	.36	.33	.31	.35	.32	.30	.29
			4	.36	.31	.28	.35	.31	.28	.34	.30	.27	.32	.29	.26	.31	.28	.26	.25
			5	.32	.27	.24	.31	.27	.24	.30	.26	.23	.29	.25	.23	.28	.25	.22	.21
			6	.29	.24	.20	.28	.24	.20	.27	.23	.20	.26	.22	.20	.25	.22	.19	.18
			7	.26	.21	.18	.25	.21	.18	.24	.20	.17	.23	.20	.17	.22	.19	.17	.16
			8	.23	.19	.16	.23	.18	.15	.22	.18	.15	.21	.18	.15	.20	.17	.15	.14
			9	.21	.17	.14	.21	.16	.14	.20	.16	.13	.19	.16	.13	.19	.15	.13	.12
			10	.19	.15	.12	.19	.15	.12	.18	.14	.12	.18	.14	.12	.17	.14	.12	.11
Porcelain-enameled reflector with 14°CW shielding (13%↑, 74%↓)	III	1.3	0	1.00	1.00	1.00	.96	.96	.96	.89	.89	.89	.82	.82	.82	.76	.76	.76	.73
			1	.88	.85	.82	.85	.82	.79	.79	.77	.74	.73	.72	.70	.68	.67	.66	.63
			2	.78	.72	.67	.75	.70	.66	.70	.66	.62	.65	.62	.59	.61	.58	.56	.53
			3	.69	.62	.57	.66	.60	.56	.62	.57	.53	.58	.54	.51	.54	.51	.48	.46
			4	.61	.54	.48	.59	.52	.47	.55	.50	.45	.52	.47	.43	.49	.45	.42	.39
			5	.54	.46	.41	.52	.45	.40	.49	.43	.39	.46	.41	.37	.43	.39	.36	.33
			6	.48	.41	.35	.47	.40	.35	.44	.38	.34	.41	.36	.32	.39	.34	.31	.29
			7	.43	.36	.31	.42	.35	.30	.40	.34	.29	.37	.32	.28	.35	.31	.27	.25
			8	.39	.32	.27	.38	.31	.26	.36	.30	.25	.34	.28	.24	.32	.27	.24	.22
			9	.35	.28	.23	.34	.27	.23	.32	.26	.22	.30	.25	.21	.28	.24	.20	.19
			10	.32	.25	.20	.31	.24	.20	.29	.23	.19	.28	.22	.19	.26	.21	.18	.17

Coefficients of Utilization for 20 Per Cent Effective Floor Cavity Reflectance (ρfc = 20)

TABLE 9.11. CONTINUED.

Typical Luminaire		Maint. Cat.	Maximum S/MH Guide[d]	RCR[c]	80			70			50			30			10			0	WDRC
	Typical Distribution and Per Cent Lamp Lumens			ρcc[a] →	50	30	10	50	30	10	50	30	10	50	30	10	50	30	10	0	
				ρw[b] →	Coefficients of Utilization for 20 Per Cent Effective Floor Cavity Reflectance (ρFC = 20)																
Diffuse aluminum reflector with 35°CW shielding	17°, 66°	II	1.5/1.3	0	.94	.94	.94	.90	.90	.90	.82	.82	.82	.75	.75	.75	.69	.69	.69	.66	
				1	.85	.82	.80	.82	.79	.77	.75	.73	.72	.69	.68	.66	.64	.63	.62	.59	.18
				2	.76	.72	.68	.74	.70	.66	.68	.65	.62	.63	.61	.58	.58	.56	.55	.52	.17
				3	.69	.63	.59	.66	.61	.57	.62	.58	.54	.57	.54	.51	.53	.51	.48	.46	.17
				4	.62	.56	.51	.60	.54	.50	.56	.51	.47	.52	.48	.45	.48	.45	.43	.41	.16
				5	.55	.49	.44	.53	.48	.43	.50	.45	.41	.47	.43	.39	.44	.40	.38	.36	.15
				6	.50	.43	.39	.48	.42	.38	.45	.40	.36	.42	.38	.35	.40	.36	.33	.31	.15
				7	.45	.38	.34	.43	.37	.33	.41	.36	.32	.38	.34	.30	.36	.32	.29	.27	.14
				8	.40	.34	.29	.39	.33	.29	.37	.31	.28	.34	.30	.26	.32	.28	.25	.24	.13
				9	.36	.30	.25	.35	.29	.25	.33	.28	.24	.31	.26	.23	.29	.25	.22	.20	.13
				10	.33	.26	.22	.32	.26	.22	.30	.25	.21	.28	.23	.20	.26	.22	.19	.18	.12
Porcelain-enameled reflector with 30°CW x 30°LW shielding	23½°, 57°	II	1.0	0	.90	.90	.90	.85	.85	.85	.76	.76	.76	.68	.68	.68	.60	.60	.60	.57	
				1	.81	.78	.76	.77	.74	.72	.69	.67	.66	.62	.61	.60	.56	.55	.54	.57	.16
				2	.72	.68	.64	.69	.65	.62	.62	.59	.57	.56	.54	.52	.51	.49	.47	.45	.16
				3	.65	.59	.55	.62	.57	.53	.56	.52	.49	.51	.48	.46	.46	.44	.42	.39	.15
				4	.58	.52	.48	.56	.50	.46	.51	.46	.43	.46	.43	.40	.42	.39	.37	.35	.14
				5	.52	.46	.41	.50	.44	.40	.46	.41	.38	.42	.38	.35	.38	.35	.33	.30	.13
				6	.47	.41	.36	.45	.39	.35	.41	.37	.33	.38	.34	.31	.35	.31	.29	.27	.13
				7	.43	.36	.32	.41	.35	.31	.38	.33	.29	.34	.30	.27	.32	.28	.26	.24	.12
				8	.38	.32	.28	.37	.31	.27	.34	.29	.26	.31	.27	.24	.29	.25	.23	.21	.11
				9	.35	.29	.24	.33	.28	.24	.31	.26	.22	.28	.24	.21	.26	.22	.20	.18	.11
				10	.32	.26	.22	.30	.25	.21	.28	.23	.20	.26	.22	.19	.24	.20	.18	.16	.10

[a] ρcc = per cent effective ceiling cavity reflectance.
[b] ρw = per cent wall reflectance.
[c] RCR = Room Cavity Ratio.
[d] Maximum S/MH guide—ratio of maximum luminaire spacing to mounting or ceiling height above work-plane.

*Source: Fig. 9.12. IES Lighting Handbook.

TABLE 9.12. MULTIPLYING FACTORS FOR OTHER THAN 20% EFFECTIVE FLOOR CAVITY REFLECTANCE

% Effective Ceiling Cavity Reflectance, ρcc	80				70				50			30			10		
% Wall Reflectance, ρw	70	50	30	10	70	50	30	10	50	30	10	50	30	10	50	30	10

For 30 Per Cent Effective Floor Cavity Reflectance (20 Per Cent = 1.00)

Room Cavity Ratio	70	50	30	10	70	50	30	10	50	30	10	50	30	10	50	30	10
1	1.092	1.082	1.075	1.068	1.077	1.070	1.064	1.059	1.049	1.044	1.040	1.028	1.026	1.023	1.012	1.010	1.008
2	1.079	1.066	1.055	1.047	1.068	1.057	1.048	1.039	1.041	1.033	1.027	1.026	1.021	1.017	1.013	1.010	1.006
3	1.070	1.054	1.042	1.033	1.061	1.048	1.037	1.028	1.034	1.027	1.020	1.024	1.017	1.012	1.014	1.009	1.005
4	1.062	1.045	1.033	1.024	1.055	1.040	1.029	1.021	1.030	1.022	1.015	1.022	1.015	1.010	1.014	1.009	1.004
5	1.056	1.038	1.026	1.018	1.050	1.034	1.024	1.015	1.027	1.018	1.012	1.020	1.013	1.008	1.014	1.009	1.004
6	1.052	1.033	1.021	1.014	1.047	1.030	1.020	1.012	1.024	1.015	1.009	1.019	1.012	1.006	1.014	1.008	1.003
7	1.047	1.029	1.018	1.011	1.043	1.026	1.017	1.009	1.022	1.013	1.007	1.018	1.010	1.005	1.014	1.008	1.003
8	1.044	1.026	1.015	1.009	1.040	1.024	1.015	1.007	1.020	1.012	1.006	1.017	1.009	1.004	1.013	1.007	1.003
9	1.040	1.024	1.014	1.007	1.037	1.022	1.014	1.006	1.019	1.011	1.005	1.016	1.009	1.004	1.013	1.007	1.002
10	1.037	1.022	1.012	1.006	1.034	1.020	1.012	1.005	1.017	1.010	1.004	1.015	1.009	1.003	1.013	1.007	1.002

For 10 Per Cent Effective Floor Cavity Reflectance (20 Per Cent = 1.00)

Room Cavity Ratio	70	50	30	10	70	50	30	10	50	30	10	50	30	10	50	30	10
1	.923	.929	.935	.940	.933	.939	.943	.948	.956	.960	.963	.973	.976	.979	.989	.991	.993
2	.931	.942	.950	.958	.940	.949	.957	.963	.962	.968	.974	.976	.980	.985	.988	.991	.995
3	.939	.951	.961	.969	.945	.957	.966	.973	.967	.975	.981	.978	.983	.988	.988	.992	.996
4	.944	.958	.969	.978	.950	.963	.973	.980	.972	.980	.986	.980	.986	.991	.987	.992	.996
5	.949	.964	.976	.983	.954	.968	.978	.985	.975	.983	.989	.981	.988	.993	.987	.992	.997
6	.953	.969	.980	.986	.958	.972	.982	.989	.977	.985	.992	.982	.989	.995	.987	.993	.997
7	.957	.973	.983	.991	.961	.975	.985	.991	.979	.987	.994	.983	.990	.996	.987	.993	.998
8	.960	.976	.986	.993	.963	.977	.987	.993	.981	.988	.995	.984	.991	.997	.987	.994	.998
9	.963	.978	.987	.994	.965	.979	.989	.994	.983	.990	.996	.985	.992	.998	.988	.994	.999
10	.965	.980	.989	.995	.967	.981	.990	.995	.984	.991	.997	.986	.993	.998	.988	.994	.999

For 0 Per Cent Effective Floor Cavity Reflectance (20 Per Cent = 1.00)

Room Cavity Ratio	70	50	30	10	70	50	30	10	50	30	10	50	30	10	50	30	10
1	.859	.870	.879	.886	.873	.884	.893	.901	.916	.923	.929	.948	.954	.960	.979	.983	.987
2	.871	.887	.903	.919	.886	.902	.916	.928	.926	.938	.949	.954	.963	.971	.978	.983	.991
3	.882	.904	.915	.942	.898	.918	.934	.947	.936	.950	.964	.958	.969	.979	.976	.984	.993
4	.893	.919	.941	.958	.908	.930	.948	.961	.945	.961	.974	.961	.974	.984	.975	.985	.994
5	.903	.931	.953	.969	.914	.939	.958	.970	.951	.967	.980	.964	.977	.988	.975	.985	.995
6	.911	.940	.961	.976	.920	.945	.965	.977	.955	.972	.985	.966	.979	.991	.975	.986	.996
7	.917	.947	.967	.981	.924	.950	.970	.982	.959	.975	.988	.968	.981	.993	.975	.987	.997
8	.922	.953	.971	.985	.929	.955	.975	.986	.963	.978	.991	.970	.983	.995	.976	.988	.998
9	.928	.958	.975	.988	.933	.959	.980	.989	.966	.980	.993	.971	.985	.996	.976	.988	.998
10	.933	.962	.979	.991	.937	.963	.983	.992	.969	.982	.995	.973	.987	.997	.977	.989	.999

Source: IES Lighting Handbook (1972).

in the following example problem. When using the Tables it is necessary to interpolate for numbers between categories in many places. This can become rather confusing. An alternate procedure would be to use the nearest available categories.

Example 9.1.

Coefficient of Utilization.

Determine the coefficient of utilization for a farm shop which is 6 m wide and 14 m long with a ceiling height of 5 m. The work plane is to be 1.5 m above the floor. Fluorescent lights (Porcelain-enameled reflector with 14° CW shielding) are suspended 1 m below the ceiling. Ceiling, wall and floor reflectances of 80%, 50% and 20% can be assumed.

Coefficient of Utilization Worksheet

Following through the CU worksheet we obtain the following results.

Room Identification   Farm Shop

Step 1: Fill in sketch.

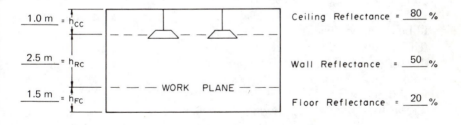

Room Dimensions: Length 14 m , Width 6 m , Height 5 m

Step 2: Determine Cavity Ratios from formula.

$$\text{Cavity Ratio} = \frac{5h \, (\text{Room Width} + \text{Room Length})}{\text{Room Width} \times \text{Room Length}}$$

where $h = h_{RC}$ for RCR

$h = h_{CC}$ for CCR

$h = h_{FC}$ for FCR

Room Cavity Ratio (RCR) = ___3.0___

Ceiling Cavity Ratio (CCR) = ___1.2___

Floor Cavity Ratio (FCR) = ___1.8___

Step 3: Obtain Effective Cavity Reflectance from Table 9.10

Effective Ceiling Cavity Reflectance (ρ_{CC}) = ___64___ %

Effective Floor Cavity Reflectance (ρ_{FC}) = ___18___ %

Step 4: Obtain Coefficient of Utilization from Table 9.11

Coefficient of Utilization (CU) = ___0.65___ .

Step 5: Adjust CU if Floor-Cavity Ratio is other than 20%, Adjustment Factor from Table 9.12.

Adjustment Factor (AF) = ___0.99___ .

Adjusted CU = CU at 20% × AF

= ___0.65___ × ___0.99___ = ___0.64___

9.5.2 Selection of Light Loss Factors

The three factors most likely to cause a decrease in illumination level over time are room surface dirt, dirt on the luminaire, and lamp lumen depreciation. Each of these factors can be considered in establishing a Light Loss Factor (LLF). The total Light Loss Factor will be the product of several factors. Determining each of the factors will be discussed briefly.

The accumulation of dirt on room surfaces reduces the amount of light reflected to the work plane. To account for this a Room Surface Dirt Depreciation (RSDD) factor is established. The factor is determined by:

1) Finding the expected dirt depreciation for the type of atmosphere and time between cleaning of the room surfaces using Fig. 9.12. For example, from the graph in Fig. 9.12, if the atmosphere is very dirty and the room surface is cleaned every 12 months, the expected dirt depreciation would be 30%.

2) Knowing the expected dirt depreciation the type of luminaire and the room cavity ratio (RCR), the Room Surface Dirt Depreciation (RSDD) factor can be obtained from the table in Fig. 9.11. For example, for a dirt-depreciation of 30% a semidirect luminaire and a Room Cavity Ratio (RCR) of 5, the RSDD would be 0.84.

The accumulation of dirt on luminaires results in loss of light out-

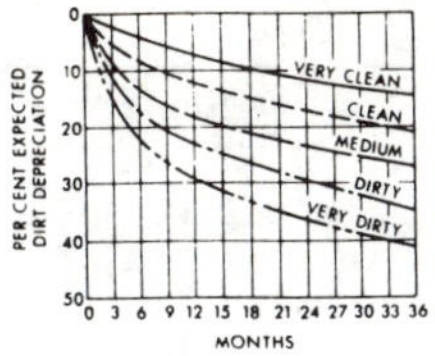

	Luminaire Distribution Type																			
	Direct				Semi-Direct				Direct-Indirect				Semi-Indirect				Indirect			
Per Cent Expected Dirt Depreciation	10	20	30	40	10	20	30	40	10	20	30	40	10	20	30	40	10	20	30	40
Room Cavity Ratio																				
1	.98	.96	.94	.92	.97	.92	.89	.84	.94	.87	.80	.76	.94	.87	.80	.73	.90	.80	.70	.60
2	.98	.96	.94	.92	.96	.92	.88	.83	.94	.87	.80	.75	.94	.87	.79	.72	.90	.80	.69	.59
3	.98	.95	.93	.90	.96	.91	.87	.82	.94	.86	.79	.74	.94	.86	.78	.71	.90	.79	.68	.58
4	.97	.95	.92	.90	.95	.90	.85	.80	.94	.86	.79	.73	.94	.86	.78	.70	.89	.78	.67	.56
5	.97	.94	.91	.89	.94	.90	.84	.79	.93	.86	.78	.72	.93	.86	.77	.69	.89	.78	.66	.55
6	.97	.94	.91	.88	.94	.89	.83	.78	.93	.85	.78	.71	.93	.85	.76	.68	.89	.77	.66	.54
7	.97	.94	.90	.87	.93	.88	.82	.77	.93	.84	.77	.70	.93	.84	.76	.68	.89	.76	.65	.53
8	.96	.93	.89	.86	.93	.87	.81	.75	.93	.84	.76	.69	.93	.84	.76	.68	.88	.76	.64	.52
9	.96	.92	.88	.85	.93	.87	.80	.74	.93	.84	.76	.68	.93	.84	.75	.67	.88	.75	.63	.51
10	.96	.92	.87	.83	.93	.86	.79	.72	.93	.84	.75	.67	.92	.83	.75	.67	.88	.75	.62	.50

Source: IES Lighting Handbook (1972).

FIG. 9.12. ROOM SURFACE DIRT DEPRECIATION FACTORS

put. The loss factor is known as the Luminaire Dirt Depreciation (LDD) factor. The LDD is determined by:

1) Finding the luminaire maintenance category, I–VI, from Table 9.11 for the luminaire to be used.

2) Selecting the appropriate set of curves for the maintenance category in Fig. 9.13. Using the type of environment and months between cleaning an LDD factor can be obtained. For example, a luminaire in maintenance category II, in a very dirty environment, and which is cleaned every 15 months would have an LDD of 0.8.

As discussed earlier, the output of a lamp will diminish over time. The type of lamp replacement program will control the lamp lumen depreciation. LLD factors from Tables 9.2, 9.3 and 9.4 or manufacturers data should be used. Seventy percent of the average rated life is the minimum reached for a situation where burnouts are promptly replaced.

Other factors, generally of less significance, can also be included for a more detailed analysis. These factors are described in Section 9 of the IES Handbook.

The total light loss factor (LLF) is the product of the individual factors. For example, for a Room Surface Dirt Depreciation (RSDD) of 0.85, a Lamp Lumen Depreciation (LLD) of 0.80 and a Luminaire Dirt Depreciation (LDD) of 0.80, the Light Loss Factor (LLF) is 0.85 × 0.80 × 0.80 = 0.54.

Once the Coefficient of Utilization and the Light Loss Factor have been established, either the area per luminaire and hence the num-

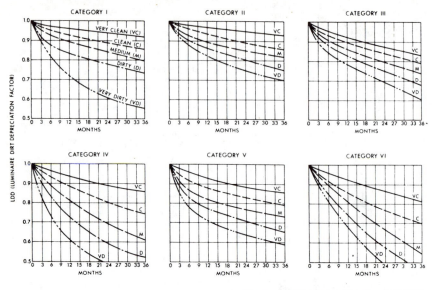

Source: *IES Lighting Handbook (1972).*

FIG. 9.13.   LUMINAIRE DEPRECIATION FACTORS (LDD)

ber of luminaires required or the maintained illumination can be calculated.

Example 9.2.

Calculating Number of Luminaires Required.

A room is to be illuminated with 150 watt incandescent lamps placed in a standard porcelain-enameled dome. If the room is 5 m by 8 m and has a Coefficient of Utilization of 0.5 and a Light Loss Factor of 0.7, how many luminaires are needed to maintain a lighting level of 200 lux?

From Table 9.2, 150 watt bulbs have an initial rating of 2880 lumens/bulb.

Calculating area per luminaire

$$\frac{\text{Area per}}{\text{Luminaire}} = \frac{\text{Lamp Lumens per Lumin.} \times \text{CU} \times \text{LLF}}{\text{Maintained Illumination}}$$

$$= \frac{2880 \times 0.5 \times 0.7}{200}$$

$$= 5.0 \ \text{m}^2/\text{lumin.}$$

$$\frac{\text{Number of}}{\text{Luminaires}} = \frac{\text{Total Area}}{\text{Area per Luminaire}} = \frac{5 \ \text{m} \times 8 \ \text{m}}{5.0 \ \text{m}^2/\text{lumin.}} = 8 \ \text{luminaires}$$

Eight luminaires are required to provide the necessary light.

9.5.3 Placement of Luminaires for Interior Lighting

In many instances the task controlling the lighting needs may be located or relocated at any point in the room, therefore uniformity in lighting level is highly desirable. Uniformity is considered acceptable if the maximum and minimum values in the room are not more than 1/6 above or below the average.

To achieve acceptable uniformity, luminaires should not be spaced too far apart or too far from the walls. The type of luminaire and height above the working plane are the main controlling factors. A guide for maximum spacing-to-mounting height (S/MH) is given for various luminaire types in Table 9.11.

The commonly used practice of letting the distance from the luminaires to the wall equal one-half of the distance between rows results in inadequate lighting near the wall. It is recommended that the first row be located within one-third the maximum spacing from

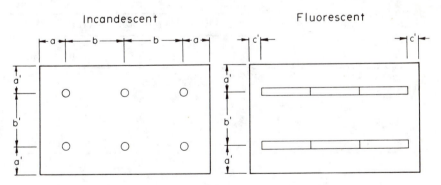

a and a' less than or equal to (max. spacing)/3.0
b and b' less than or equal to max. spacing
c less than or equal 0.6 m

FIG. 9.14. LUMINAIRE PLACEMENT LIMITS

the wall. To prevent poor illumination at the ends of rooms, rows of fluorescent lamps should come within 0.6 m (2 ft) of the wall before ending. These recommendations are summarized in Fig. 9.14.

9.6 OUTDOOR FLOODLIGHTING

A common method for floodlight calculations is the beam-lumen method. The beam-lumen method is quite similar to the method of interior lighting, however, it must take into consideration the fact that floodlights are not usually directly above the surface, but instead are aimed at various angles to the surface.

The basic formula for floodlighting from one light is:

$$\text{Average Maintained Illumination} = \frac{\text{Beam Lumens} \times \text{Coefficient of Beam Utilization} \times \text{Lightloss Factor}}{\text{Area}}$$

Beam lumens are equal to lamp lumens multiplied by the beam efficiency of the floodlight. Outdoor floodlight luminaire designations and minimum beam efficiencies are given in Table 9.13.

TABLE 9.13. OUTDOOR FLOODLIGHT LUMINAIRE DESIGNATIONS

Beam Spread Degrees	NEMA Type	Minimum Efficiencies (%)				
		Incandescent Lamps		Mercury Lamps		Fluorescent Lamps
		Effective reflector Area (m^2)				
		Under 0.146	Over 0.146	Under 0.146	Over 0.146	Any
10 up to 18	1	34	35	—	—	20
18 up to 29	2	36	36	22	30	25
29 up to 46	3	39	45	24	34	35
46 up to 70	4	42	50	35	38	42
70 up to 100	5	46	50	38	42	50
100 up to 130	6	—	—	42	46	55
130 and up	7	—	—	46	50	55

Source: Anon. (1964).

Coefficient of Beam Utilization (CBU) is a ratio of the lumens falling on the area to be illuminated to the total beam lumens. If the floodlight is being used from an elevated location to illuminate a ground area, the CBU would be very close to one.

$$CBU = \frac{\text{Lumens falling in Area}}{\text{Total Beam Lumens}}$$

The light loss factor (LLF) allows for depreciation of the lamp output over time and depreciation due to collection of dirt. Depreciation of the lamp will depend on the type and size of lamp, while depreciation for dirt will depend on the type of enclosure. The total factor generally varies from 0.65 to 0.85.

Example 9.3.

Calculating Floodlighting.

What would be the average illumination level from a 250 watt mercury lamp mounted at an elevation of 8 m if the lamp has a beam spread of 120° in a circular pattern and the reflector area is 0.2 m²? Sketch of area illuminated.

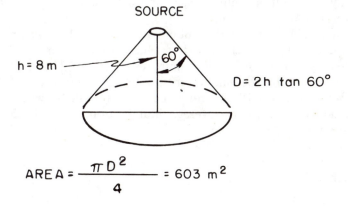

SOURCE

$h = 8\,m$

$D = 2h\,\tan 60°$

$$AREA = \frac{\pi D^2}{4} = 603\ m^2$$

From Table 9.4 — 250 W Mercury Lamp = 11,270 lm
From Table 9.13 — Beam Efficiency = 46%
Assuming CBU ≃ 1.0 and LLF ≃ 75%

$$\text{Average Maintained} = \frac{\text{Beam Lumens} \times \text{CBU} \times \text{LLF}}{\text{Area}}$$

$$\text{Illumination} = \frac{11,270 \times 0.46 \times 0.75}{603} = 6.4\ lx$$

EXERCISES

1. What is the initial lumen output of 5-150 watt incandescent bulbs? What would be their output at 70% of rated life?
2. How many 40 watt cool white fluorescent lamps would be required to have an initial lamp output of 9600 lumens?
3. Generate a graph of lumens per watt vs. bulb wattage for:

   (a) incandescent lamps
   (b) fluorescent lamps
   (c) mercury vapor lamps

4. Design a general lighting system in a milking parlor using recessed luminaires with incandescent lamps. The milk house is to have the walls washed every 6 months. The interior is 12 m by 14 m with 3.5 m ceiling.
5. Design a general lighting system for the milking parlor in the previous problem using fluorescent lighting.
6. Design a lighting system for a livestock housing area which is 10 m by 20 m with a ceiling height of 4.5 m. Assume luminaires will be mounted at the ceiling.
7. What would be the average illumination from a 175 watt mercury vapor floodlight, if the floodlight has a reflector area of 0.1 m^2, a beam spread of 90°, and illuminates an area of 215 m^2?
8. What would be the maximum mounting height for a 400 watt mercury vapor (.16 m^2 reflector area) floodlight with a beam spread of 120° if an average illumination level of 25 lux is desired?

REFERENCES

ANON. 1970. Practices for Industrial Lighting. American National Standards Institute: Allol, Published by Illuminating Engineers Society, N.Y., Publication RP-7.

ANON. 1978. Lighting for Dairy Farms and the Poultry Industry. Agricultural Engineers' Yearbook. Am. Soc. of Agric. Eng., St. Joseph, Mo.

KAUFMAN, J. E. and CHRISTENSEN, J. F. 1972. IES Lighting Handbook. Fifth Edition. Illuminating Eng. Soc., N.Y.

10

Electric Heating

10.1 FEATURES OF ELECTRIC HEATING

Today electricity enables us to produce heat in nearly any desired location and to control this heat with extreme accuracy. There are several features of electric heating which have lead to its extensive use in agriculture, other industries and residences.

Ease and accuracy of control is a primary feature. Temperature can be controlled within close tolerances with simple controls, partially due to the almost instantaneous response to controls. Electric heating does not involve combustion, therefore safety is improved due to the absence of combustion, combustible materials and the products of combustion. Electric heaters can be designed to distribute heat evenly over a large area or localize heating as desired. Equipment involved is compact, quiet, clean, relatively inexpensive, may be portable, and may be installed in remote or concealed locations. All these features have lead to extensive use of electric heat.

Electric energy can be transformed into heat by any one or a combination of the three basic components of electrical impedance: resistance, capacitance, and inductance. The methods can be classified as: 1) resistance heating, 2) dielectric heating, 3) induction heating, and 4) electric-arc heating. Each of the four methods will be discussed briefly in the following sections.

10.2 RESISTANCE HEATING

Resistance heating is probably the most common method of producing heat from electricity. As current flows through a resistance, the electrical energy is changed to heat. This is the principle of operation for most heating appliances, such as stoves, ovens, heat lamps, and area heaters. Heat is transferred by one or a combination of the three heat-transfer methods: conduction, convection, and radiation. The electric stove transfers heat to a pan by contact

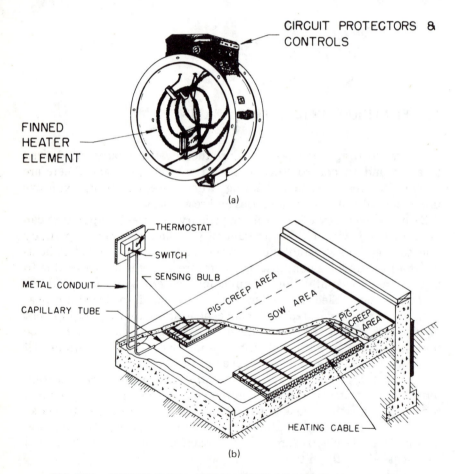

(a)

(b)

FIG. 10.1. AGRICULTURAL APPLICATIONS OF RESISTANCE HEATING
(a) CROP DRYING AIR HEATER
(b) FLOOR HEAT FOR SWINE FARROWING

(conduction) with the stove elements. A soldering iron is another sample of conduction heating. The heat is conducted from the element to the copper tip and then to the metal to be soldered.

For convection, a medium such as air or water conveys the heat to the desired location. A fan-type heater unit is an example of this type heating. The air near the resistance element is heated and the fan then moves the air and heat to the space to be heated.

Radiation is the third basic method of heat transfer. The heat lamp which transfers heat at infrared wavelengths is an example of radiant electric heating.

The energy converted to heat in a resistance is expressed as,

$$P = EI = \frac{E^2}{R} = I^2 R$$

where   P = power in watts

I = current flow in amperes

R = resistance of heating unit at operating temperature

E = voltage drop across the resistance.

A material to be used as a resistive heating element should have the following properties:

(1) high electrical resistance,
(2) low temperature coefficient of expansion to avoid mechanical stresses in the material,
(3) high resistance to oxidation to avoid deterioration during use,
(4) high melting point,
(5) high mechanical strength, and
(6) reasonable cost.

Pure metals generally do not satisfy these criteria. Alloys such as nickel-chromium combinations are most generally used.

10.3 DIELECTRIC HEATING

If we rapidly bend a piece of wire back and forth, eventually it will break. If we touch the wire near where it broke, it is likely to feel hot. This heat has been created by the stresses we placed on the material by bending it. Similar heat is created in a poor conductor used as a dielectric material in a capacitor. Application of a high-

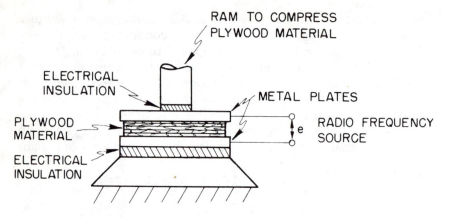

FIG. 10.2.   DIELECTRIC HEATING FOR PLYWOOD PRODUCTION

frequency source across the capacitor will cause electrons to oscillate within the material and create heat through their motion.

The dielectric process is sometimes used in the manufacture of plywood. The separate sheets of wood, with the glue between them are pressed between the plates. The stresses produced in the wood produce heat to cure the glue. Dielectric heating is also used for tasks such as curing rubber, drying textiles, and heat treatment of plastics.

The power converted to heat in a dielectric heating process can be expressed as,

$$P = 1.77 \ f \left(\frac{e}{t}\right)^2 K \cos \theta \times 10^{-9} \ \frac{w}{m^3}$$

where

f   = frequency in Hertz

e   = applied voltage

t   = thickness of charge in m

K = dielectric constant of material

$\cos \theta$ = power factor.

Frequencies used in dielectric heating are normally in the range of 10 to 30 megahertz.

The efficiency of the heating depends upon the type, size, and shape of the material being heated and upon the frequency and voltage of the electric source. The efficiency of dielectric heating is normally about 50%.

10.4 INDUCTION HEATING

For induction heating, a conducting material to be heated is exposed to a high-frequency electromagnetic field. The conductor in the magnetic field becomes heated due to induced eddy currents. Eddy currents are set up when a rapidly changing magnetic field cuts through a metallic object. The high currents flowing inside the metal cause it to heat. Because of skin effects, eddy currents are greatest in magnitude near the surface. The degree of concentration of current near the surface is controlled by the frequency, therefore depth of heating can be controlled by varying the frequency of the source.

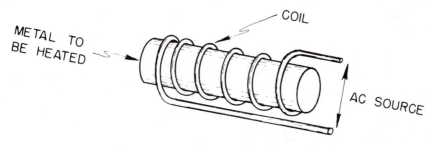

FIG. 10.3.   INDUCTION HEATING OF A ROD

The power or heat generated by induction heating can be expressed as,

$$P = 1.95 \left(\frac{\pi N}{R}\right)^2 I^2 S(ufr)^{1/2} \times 10^{-2} \text{ W}$$

where

N = number of turns in coil

L = length of charge-m

I = current in coil-amperes

S = shape factor of charge

u = permeability of charge

f = frequency in Hz

r = resistivity of charge in ohm $\times$ cm

The shape factor is dependent on the dimensions of the material to be heated, the electrical properties of the material, and the source frequency. The efficiency of induction heating is often near 50%.

Induction heating is commonly used for hardening and annealing of steel.

10.5 ARC HEATING

Arc heating is commonly used in the production of steel and in arc welding. Arc heating is really another form of resistance heating.

For arc welding, an arc passes through a column of vaporized metal. This very large current in a small area melts the metal at the ends of the arc. Filler metal from the electrode is deposited on the molten

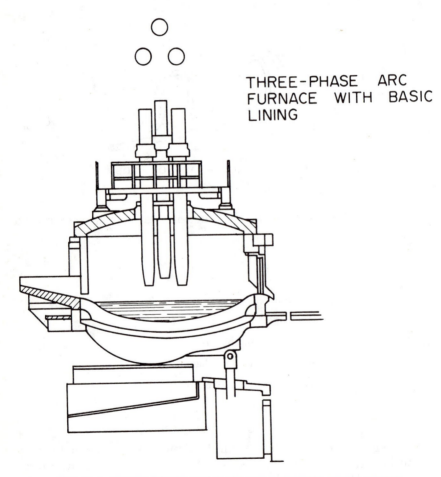

THREE-PHASE ARC FURNACE WITH BASIC LINING

FIG. 10.4.   THREE-PHASE ARC FURNACE FOR STEEL PRODUCTION

base metal. The "penetration" of an arc weld is the depth to which the base metal is melted.

Electric-arc furnaces for alloy production can utilize either three-phase or single-phase AC sources. An arc is maintained between either a carbon or graphite electrodes and the material to be melted. The efficiency of such a furnace is in the range of 75 to 90% with power factors of 0.7 to 0.9.

REFERENCES

BAUMEISTER, T. 1958. Mark's Standard Handbook for Mechanical Engineers, McGraw-Hill, N.Y.

RICHEY, C. B., JACOBSON, P. and HALL, C. W. 1961. Agricultural Engineers' Handbook, McGraw-Hill, N.Y.

Standby Power Units

11.1 PURPOSES AND IMPORTANCE

Electricity plays a critical role in farm operations. However, the importance of a continuous electrical supply is not always recognized until a power outage occurs. Some operations such as ventilation of confinement poultry or swine operations may only be able to tolerate very brief power outages before the animals begin to suffocate. Operation of water pumps, mechanical feeders, milk coolers, and milking machines are all dependent on electrical energy. In the home, loss of power may mean loss of refrigeration and possible spoilage of food or the loss of heat due to inactive furnace controls. The ever-increasing dependence on a continuous supply of electrical energy has created interest and need for standby electrical generation equipment.

Standby power equipment can eliminate the inconveniences, frustrations and economic risks of power interruptions. A farm operator must decide if he will buy the "insurance" of a standby generator or if he will accept the risk of power failure. The cost of the system must be compared to the possible financial loss and inconvenience created by a power outage.

11.2 STANDBY GENERATOR TYPES

Standby generating equipment can be divided into two general classes, stationary and portable. Stationary units generally are made up of a generator with a motor coupled to the generator. Portable units may be either driven by an engine fixed to the generator or by the PTO drive of a tractor. Portable units are generally smaller than stationary units, however they can be more easily moved. Portable units can also be used for operating tools such as welders, saws, drills, or lights for field work or repair jobs in remote locations where power is not otherwise available.

FIG. 11.1. ENGINE DRIVEN STATIONARY STANDBY GEN-ERATOR

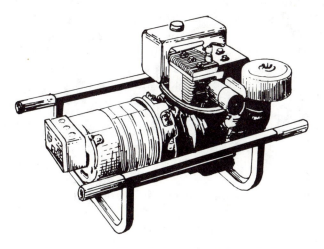

FIG. 11.2. PORTABLE STANDBY GENERATOR

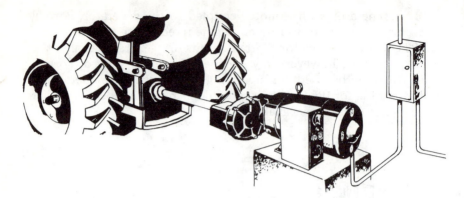

FIG. 11.3. PTO DRIVEN STANDBY GENERATOR

Stationary units are usually fairly large units (20 to 80 kilowatts). They may be either manually or automatically controlled. Situations, such as some confinement poultry operations where loss of power even for a few minutes is critical, can be protected by a unit which automatically starts when the normal power fails. Gasoline, LP gas, and diesel engines are available to power these types of units. LP fuels are generally cleaner burning than gasoline. This tends to promote longer engine life and reduced frequency of service. Gasoline has the additional disadvantage that it tends to form a varnish in the system during storage which may hinder starting. Diesel units are high in initial cost except for very large engines. However, operating and maintenance costs are low for diesel engines.

Portable engine-driven units are commonly used when the electric load is small. Some examples might be for running electric power tools or lighting during field repair of equipment. Portable engine-driven units are usually powered by air-cooled gasoline engines. Units are generally in the one to four kilowatt range.

PTO driven units are the most common on farmsteads since the farm tractor is generally available to act as the drive component. PTO units may be bolted to a stationary platform or mounted on a trailer. Trailor mounted units provide flexibility in using the unit for tasks other than power failures, such as welding in remote locations. These units generally range in size from 7 to 40 kilowatts. A minimum engine power of two horsepower per one kilowatt of electric power is required to properly run the generator. For example, a 15 kilowatt generator would require a minimum of a 30 horsepower tractor to operate properly.

11.3 SIZING A STANDBY UNIT

The size generator needed for a standby unit will depend on what loads are to be operated by the generator. Systems can be designed to operate all of the loads normally found on the farm. However, systems operating part of the load are less expensive and generally more practical for farm use. When determining the size of unit needed, the first step is to make a listing of all necessary equipment to be used, such as water pumps, heaters, refrigerators, ventilation fans, other motors, etc. When possible, list the nameplate power rating for each device. If nameplates are not accessible, Table 11.1

TABLE 11.1. ESTIMATED LOAD FOR NECESSARY ITEMS

Home		Farm		
Equipment	Watts	Equipment	Watts	Horsepower
Total Lights	3 watts/ft^2	Total Lights	1/2 watt/ft^2	—
Refrigerator	400	Milking Machine		1/2 to 5
Freezer	400 to 600	Milk Cooler		1/2 to 12
Electric Range	3,000 to 10,000	Water Pump		1/2 to 2
Clothes Dryer	4,000	Space Heater	1,000 to 5,000	—
Furnace Blower	400	Ventilation Fans		1/6 to 1/2 each
Water Heater	1,000 to 5,000	Silo Unloader		2 to 7 1/2
Washing Machine	400 to 600	Feed Grinder		1 to 7 1/2
		Feed Mixer		1/2 to 1
		Yard Light	100 to 500	—
		Shop Tools		1/6 to 1

may be helpful in determining farm and home loads. The generator need not be sized to supply the total connected load if heavy loads, such as electric ranges, water heaters, large motors, and milk coolers, can be staggered or rationed during the power outage. One main consideration is motor starting requirements. Motors can require three, four or more times the full-load current for starting. If only one large motor is involved at any one time, starting capacity for the motor may determine the size of unit needed. After the motor is started, the remaining capacity may be adequate to supply the other necessary loads. Table 11.2 gives starting and running requirements for common motors.

When calculating the generator wattage required, the starting load and running load for each motor should be considered. The best procedure is to start the largest motor first. Example 11.1 demonstrates this process.

TABLE 11.2. STARTING AND RUNNING REQUIREMENTS FOR COMMONLY USED 60-CYCLE, SINGLE-PHASE MOTORS

Motor HP rating	Approx. amps (full load) 120 V	240 V	Watts required to start Split phase	Cap. start[1]	Watts required to run (full load)
1/6	4.4		860		215
1/4	5.8		1,500	1,200	300
1/3	7.2		2,000	1,600	400
1/2	9.8	4.9		2,300	575
3/4	13.8	6.9		3,345	835
1		8		4,000	1,000
1 1/2		10		6,000	1,500
2		12		8,000	2,000
3		17		12,000	3,000
5		28		18,000	4,500
7 1/2		40	15,100[2] to	28,000	7,000
10		50	81,900[2] to	36,000	9,000

[1] Reduce 25% for repulsion induction motors.
[2] Soft start motors.

Example 11.1.

Sizing a Standby Unit.

The following loads might be considered necessary for a dairy: silo unloaders — 5 horsepower, milk cooler — 3 horsepower, vacuum pump — 1 horsepower, ventilation fan — 1/2 horsepower, and lights — 750 watts. In this example we will assume all motors can be manually controlled. To minimize the size generator needed, we can assume we will only start the silo unloader if no other loads are in use. This means we should consider two cases when sizing the units: 1) when the silo unloader is operating, and 2) when the milking operation is underway. The following listing helps determine the starting and running loads as loads are added to the system. The largest load situation will control the size of generator needed.

Load in Sequence of Starting	Size	Starting Watts	Load Sequence on Generator Starting	Running
		Situation 1 - Silo Unloader Only		
Silo Unloader	5 hp	18,000 W	18,000 W	4,500 W
		Situation 2 - Milk Operation		
Milk Cooler	3 hp	12,000 W	12,000 W	3,000 W
Vacuum Pump	1 hp	4,000 W	7,000 W	4,000 W
Ventilation Fan	1/2 hp	2,300 W	6,300 W	4,575 W
Lights	750 W			5,325 W

For Situation 2, starting the milk cooler would momentarily require 12,000 watts output from the generator. However, once started, the load would drop to 3000 watts. If the water pump is then started, the load would momentarily rise to 7000 watts and then drop to 4000 watts after the pump has been started. The ventilation fan would raise the load to 6300 watts for starting, then it would drop to 4575 watts. Lights totalling 750 watts would bring the total load to 5325 watts.

In this example the limiting factor is the starting current of the silo unloader. An 18 kilowatt generator capacity would be needed. Note that if an 18 kilowatt generator is used once the silo unloader is started, all other loads could be operated on the capacity which remains.

Some generators are rated for a certain running load and a temporary allowable overload percentage. For this example, a 12 kilowatt generator with a 50% overload capacity would also be adequate.

Most farms require a minimum of a 12 to 15 kilowatt unit. It is best to make sure the unit will not be too small. The cost of a larger unit may not be great enough to justify the purchase of a smaller unit. It is wise to check your calculations with your power supplier and generator dealer.

11.4 INSTALLATION

Wiring and equipment must be installed in accordance with the National Electrical Code (or any local code which may prevail) and the requirements of the power supplier. In particular, the National Electric Code requires that a standby unit be connected so as to prevent the inadvertent interconnection of the standby generator and the normal power supply. A double-pole, double-throw switch (Fig. 11.4) is usually installed between the power supplier's meter and the service entrance. The use of a double-throw switch prevents power from feeding back into the power supplier's line and endangering the lives of linemen doing repair work. It also prevents the main power source from feeding into the generator and possibly "burning out" the generator when power is restored. Most standby equipment guarantees are void if the transfer switch is not used.

The standby generator should be located within 7.6 m (25 ft) of the transfer switch if possible, preferably within sight. The equipment should be convenient so that no time is lost in an emergency.

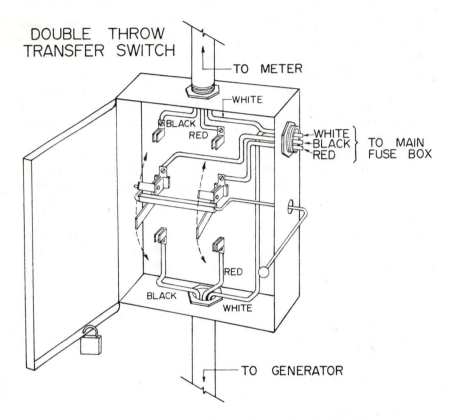

DOUBLE THROW
TRANSFER SWITCH

TO METER

WHITE

BLACK
RED

WHITE
BLACK } TO MAIN
RED   } FUSE BOX

RED

BLACK
WHITE

TO GENERATOR

FIG. 11.4. DOUBLE-POLE, DOUBLE-THROW TRANSFER SWITCH

11.5 MAINTENANCE AND OPERATION

Providing shelter for the generator and tractor during operation
may be highly advisable. Remember the unit is most likely to be
operating when the most adverse weather conditions exist. Engine
exhaust gases and heat from the engine will dictate ventilation requirements. Do not store flammable material in the area.

The standby unit should be kept clean and in good running condition at all times. Accumulation of dust and dirt may cause the unit
to overheat when operating. The unit should not be covered with a
tarp. This allows moisture to condense inside the unit causing rust.
The unit should be operated under load at least once every three
months (or as the manufacturer recommends) to be sure it is functioning properly.

For a manually operated system, the following sequence of operation should be followed:

(1) Turn off or disconnect all electrical loads on the farm.
(2) Start the standby unit.
(3) Energize the load system by engaging the transfer switch.
(4) Gradually increase the load on the generator, making periodic checks on the voltage indicator. If the voltage falls below 200 volts on a 240 volt system or below 100 volts on a 120 volt system, reduce the load by turning off some equipment. Remember large motor loads should be started first due to large starting power requirements.
(5) When commercial power is restored, return the transfer switch to the normal position. Do not stop the standby unit abruptly, rather allow the unit to roll to a stop freely. Sudden stops can damage the unit.

EXERCISES

What type and size standby unit and what operating procedures would be needed for the following situations?

1. Beef feeder operation with necessary loads of 2 silo unloaders — 5 hp each, 1 bunk auger — 1 1/2 hp, 750 W lights, 2 manure pit ventilation fans — 1/2 hp each, 3 building ventilation fans — 1/4 hp each, and water pump — 1/2 hp.
2. Confinement poultry operation with necessary loads 10 ventilation fans — 1/4 hp each, water pump — 1 hp, 2 mechanical feeders — 1/2 hp each, 8,000 W of lighting load.
3. Dairy operation with milking equipment — 1 hp, milk pipeline pump — 1/2 hp, water pump — 1 hp, milk cooler — 3 hp, lights — 600 W, gutter cleaner — 3 hp.

REFERENCES

ANON. 1977. National Electric Code 1978. Natl. Fire Protection Assoc., Boston.
ASAE. 1977. Installation and maintenance of farm standby electric power. Agricultural Engineers Yearbook. Am. Soc. of Agric. Eng., St. Joseph, Mo.

CAMPBELL, L. E. 1960. Standby Electric Power Equipment. USDA Leaflet 480. Washington, D.C.
LUNDSTROM, D. R. and KARSKY, R. 1972. Electric standby power. Electric Farm Power. Agric. Eng. Dept., N. D. St. Univ., Fargo, North Dakota.
McCURDY, J. A. Standby Equipment for Electric Power Interruptions. Extension Publications, Penn. St. Univ., University Park, Penn.

12

Lightning and
Lightning Protection

Lightning is one of nature's most destructive forces. This visible electrical discharge is particularly hazardous on the farm. Lightning is the largest single cause of fire in rural areas. In addition, one study indicated that lightning caused more than 80% of all accidental losses of livestock. Yet lightning-caused losses of homes and buildings can be prevented and a majority of livestock losses to lightning could be eliminated. The following sections will describe the basic nature and action of lightning and consider the measures needed to protect buildings, livestock and people from lightning.

12.1 THE NATURE OF LIGHTNING

Electrical charges are always present in the air. However, when a thunderstorm forms, it can disturb the normal balance between positive and negative charges. Air currents and temperature differences can separate positive and negative charges. Negative charges usually accumulate on the lower portion of clouds while positive charges accumulate on the ground.

The following sequence of events describes the development of a common lightning streak. When the cloud and the ground attain a potential difference of several thousand volts per centimeter, initially a negatively charged streamer travels downward in a jumpy, exploratory, irregular path. At the same time, a shorter leader may move up

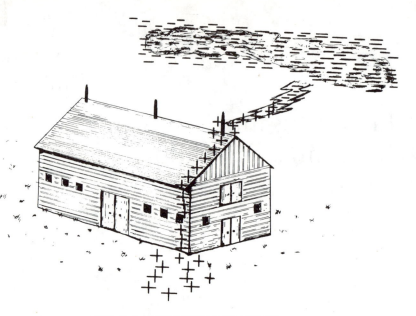

FIG. 12.1.   LIGHTNING FORMATION

from the earth. This leader generally initiates from an object such as a tree or a building because it represents the shortest (easiest) path between the ground and the cloud. When the two leaders meet or when the upper leader makes contact with the earth, the main lightning discharge surges from the earth carrying a positive charge to the cloud. The duration of the charge transfer, called a stroke, is only a few millionths of a second with peak currents of 20,000 to 150,000 amperes. Visible light is radiated for a few hundredths of a second. Normally a number of strokes will occur along the same path. The entire process is a combination of two to forty of the discharges described along the same path during a time of about one second. The discharge path averages 1.5 to 3 km. Most of the energy in lightning is dissipated as heat and light. It is the intense heat that gives rise to the thunder. In the path, molecules are separated into atoms and atoms are separated into electrons and ions.

The energy per flash of lightning is in the range of about 100,000,000 watt-seconds. This corresponds to a billion kilowatts (1.3 billion horsepower) lasting 100 microseconds. However, this is equivalent to only about 27 kilowatt hours, and hence, is less than one dollar's worth of electricity in terms of kilowatt hours of energy you purchase. We still must be awed by the tremendous forces that can be developed within a time that is practically instantaneous.

Lone trees and farm buildings are prime targets for lightning. They provide a path for the ground charges to climb nearer the charged clouds, therefore it is important we consider what is necessary to protect these items from lightning damage.

12.2 PRINCIPLES OF LIGHTNING PROTECTION

Correctly designed lightning protection systems must do two things. First, they must provide a direct easy path for the lightning to follow to the ground. Second, they must prevent damage, injury or death as the bolt travels the path. All points that the bolt is likely to strike, and all objects to which current might likely side flash must be protected.

A lightning bolt will generally follow a metallic path to ground (and from the ground up) if one is provided. However, at points along the path, the current may side flash to objects such as appliances, water lines, electric wiring, or even a person or an animal. The protection system must protect from damage due to side flashes.

A proper protection system must be designed to include all places lightning is likely to strike, to protect all objects to which current might side flash, to provide as straight a route to ground as possible, and to safely dissipate the charges.

12.3 PROTECTION FOR BUILDINGS

Lightning is likely to enter a house or farm building in one of four ways: (1) by striking a metal object like a TV antenna, cupola, or a metal track extending from the building, (2) by a direct stroke to the building, (3) by hitting a nearby tree or tall object, such as a silo, and leaping over to the building to find an easier path to ground, or (4) by striking and following a power line or an ungrounded wire fence.

Under the lightning protection code LPI-175 published by the Lightning Protection Institute, six elements are required for a building protection system. The elements are: (1) air terminals on the roof and other high points, (2) main conductors connecting air terminals and leading to grounds, (3) branch conductors to metal bodies, (4) ground electrodes and common grounding with service grounds, (5) secondary lightning arresters, and (6) annual inspection and maintenance procedure. Each of these elements will be briefly discussed. Standards for quality lightning protection also include requirements developed in Underwriters' Laboratories Requirements for Master Label for Lightning Protection UL96A, National Fire Protection Association Lightning Protection Code NFPA 78, and ASAE Engineering Practice ASAE EP 381.

Air terminals are the topmost elements of the lightning protection system and are designed to provide a safe contact point for the lightning. They must be placed and spaced such that it is assured that lightning will strike the terminals — not a vulnerable part of the building or object being protected.

Air terminals (Fig. 12.2) are made of solid copper or aluminum

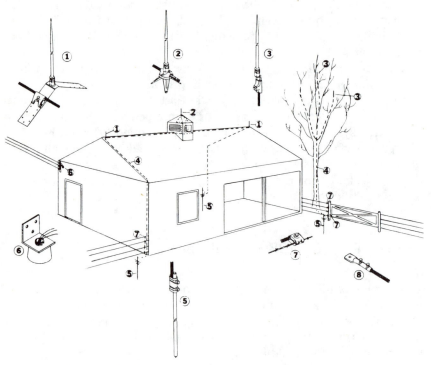

1. standard air terminal
2. air terminal for cupolas, chimneys, etc.
3. air terminal for trees
4. conductors
5. ground connectors
6. surge arrestor on incoming power lines
7. connector for wire fence
8. branch connector for eave troughs, tracks, etc.

FIG. 12.2. COMPONENTS OF A LIGHTNING PROTECTION SYSTEM

drawn to a sharp point, stand 254 mm (10 in.) to 610 mm (24 in.) high, are generally spaced at a maximum of 6.1 m (20 ft) and within 0.6 m (2 ft) of ridge ends. The main conductor should have at least 3.8 cm (1.5 in.) of contact with the air terminal. Size, anchor method, spacing requirements vary according to size and type of roof (See LPI – 175).

Conductors used for the system are divided into three categories: main, down, and branch conductors. Special copper or aluminum cables, sized according to the height of the building, are used. All conductors must be routed such that no bends with less than 20 cm (8 in.) radius or bends of greater than 90° occur.

Main conductors interconnect all air terminals and down conductors. Down conductors connect the main conductors to the ground. Down conductors should be distributed about the building at the corners. For buildings longer than 30 m (100 ft) intermediate downleads may be needed. At least two properly grounded down conductors are required on any kind of structure. Structures with more than 61 m (200 ft) of perimeter should have one more down conductor for each 30 m (100 ft) of perimeter.

Branch conductors provide protection from side flashes by bonding of metallic objects to the grounding system. Branch conductors should be connected to metal bodies such as gutters, plumbing stacks, antennas, water pipes, air conditioning units, ventilation fans, stalls and stanchions.

Lightning protection systems can be installed as concealed or semi-concealed systems. Concealed systems are run inside the building's framing. The only visible parts are the air terminal points. Concealed systems are generally installed during construction of the building. Semiconcealed systems are installed on the outside of existing buildings. Semiconcealed systems are placed behind downspouts and through building parts to conceal them as much as possible.

Grounding connections are critical to the proper functioning of the whole system. At least two ground connections are needed for every system. It is best to place them as far apart as possible, preferably on opposite corners of the building.

Grounding connections are made in one of four ways. For soils of ordinary dampness and conductivity, a 13 mm (0.5 in.) copperclad steel rod having a minimum depth of penetration of 3m (10 ft) with the top of the rod at least 0.3m (1 ft) below grade and 0.6 m (2 ft) from the building is used. The down conductors must be securely clamped to the rod. In dry, rocky, or lower conductivity soils, alternate forms of grounding include stranding copper conductor cable and burying it in a trench, clamping the down conductor to a buried sheet metal plate or clamping the down conductor to a metal water pipe.

Lightning arresters are devices which prevent dangerous surges of electricity into a building's wiring system when lightning strikes the power lines. Such arresters should be installed on both overhead and underground services at the outside electric service entrance, or at the interior service entrance, depending on local regulations. Their purpose is to supply a path to ground for lightning following the electrical supply lines, such that equipment and wiring within the building is not damaged by a lightning-caused power surge. Each arrester should be connected to the grounding system of the building's lightning protection system.

Lightning arresters should also be installed on television and radio antennas and as surge protection on major electric motors.

Common grounding has been recognized as being the most effective method of eliminating side flashes resulting from lightning discharge. Therefore all grounding mediums should be connected together. This means electric and telephone service grounds and other underground metallic piping systems (water, gas, underground conduits) should be bonded together.

12.4 PROTECTION OF TREES

Trees may be ruined or severely damaged by lightning. A well protected tree protects adjacent buildings from being damaged by a falling tree and also protects livestock or humans that may be under the tree during a storm.

Favorite lightning targets are lone trees, the tallest tree in a group, and a tall tree at the end or edge of a grove nearest the approaching storm. Trees which should be protected are those within 3 m (10 ft) of any building, those under which livestock are likely to congregate during a storm, and those that are valuable in themselves. For a grove of trees, only a few of the tallest trees need to be protected.

A tree protection system (Fig. 12.3) contains four main parts — air terminals, conductor cables, copper fasteners, and adequate grounding.

The main air terminal is placed as high as it can be securely fastened and smaller terminals are fastened on main branches. Trees of more than 1 m (3 ft) in diameter need two down conductors on opposite sides of the trunk. A suitable casing may be needed to protect the down conductors from damage by livestock. To make

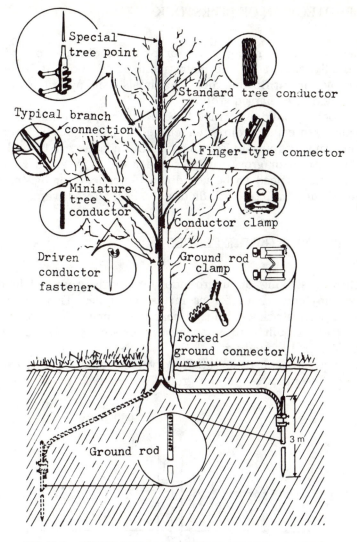

FIG. 12.3. COMPONENTS OF A TREE PROTECTION SYSTEM

ground connections, the conductors should be buried in a trench extending away from the tree at least 3.7 m (12 ft) or to the extremity of the overhanging branches. The trench needs to be shallow near the tree to prevent damage to the roots of the tree. One of the grounding methods discussed in the previous section is used at a point beyond the root spread.

12.5 PROTECTION OF LIVESTOCK

When lightning striking a tree or building kills livestock, death is not usually caused by the mainstream of the bolt. It is normally caused by current flowing through animals standing or lying in direct contact with the charged ground.

Livestock in buildings which have lightning protection systems are generally protected from such currents. The lightning protection system drains off current before dangerous charges can be built up.

Isolated trees or groves of trees under which livestock congregate during a storm need to be cut down, fenced off or lightning protected. Value of the tree will dictate the best alternative.

12.6 GROUNDING WIRE FENCES

Lightning striking on ungrounded or poorly-grounded fence can follow the fence wires for as far as two miles. This can create a hazard for livestock or humans near the fence. To prevent such a hazard, fences built with wooden posts or steel posts set in concrete need additional grounding.

Two ways to construct satisfactory grounding are 1) to use galvanized steel posts set in the ground at intervals of not more than

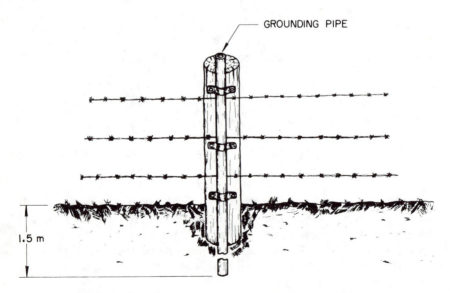

FIG. 12.4.   FENCE GROUNDING WITH A DRIVE STEEL ROD.

46 m (150 ft), or 2) to drive a 13 mm or 19 mm (1/2 or 3/4 in) steel rod or pipe 1.5 m (5 ft) into the ground along side wooden posts (Figure 12.4) at the same intervals. The rod should be fastened such that it touches all the fence wires.

12.7 PERSONAL SAFETY

The following is a list of precautions that help minimize the risk of exposure to the hazards of lightning.

If within a building: stay away from metal objects which extend to outdoor exposures such as televisions, telephone, lamp and lamp cords, water faucets and appliances; stay away from fireplace, stove, stovepipe, or chimney. Chimneys are often hit by lightning and soot deposits inside them can become the path for a stroke.

If caught outdoors: seek a low spot such as a ditch, and stay away from lone trees, wire fences, and small unprotected buildings; remain in a closed truck or automobile; do not ride in open vehicles or on horseback.

12.8 INSTALLING LIGHTNING PROTECTION

The best way to assure a complete quality installation is to be sure the work is performed or at least supervised by competent experienced personnel. Lightning protection design and installation is not a do-it-yourself job. Companies or electricians who specialize in lightning protection are widely available.

If equipment used carries the U/L "Master Label" and your system has been installed by authorized personnel, Underwriters' Laboratory will then authorize a "Master Label" plate which can be affixed to the building.

REFERENCES

ANON. 1963. Master Labeled Lightning Protection Systems, Installation Requirements, UL96A. Underwriters' Laboratory, Chicago.

ANON. 1975. Installation Code LPI - 175. Lightning Protection Institute, Harvard, Ill.

ANON. 1978. Lightning Protection Code NFPA 78. National Fire Protection Association. Boston.

ASAE. 1978. Specifications for lightning protection. Agricultural Engineers' Yearbook. Am. Soc. of Agric. Eng., St. Joseph, Mo.

GERHARDT, J. P. 1970. Lightning. In Encyclopedia International. Grolier, N.Y.

TIMMINS, M. S. 1968. Lightning Protection for the Farm. Farmer's Bulletin No. 2130. USDA, Washington, D.C.

TOWNE, H. M. 1956. Lightning — Its Behavior and What to Do About It. United Lightning Protection Assoc., Ithaca, N.Y.

Solar and Wind Energy Sources for Electricity

13.1 INTRODUCTION TO SOLAR ENERGY

A number of different methods, both direct and indirect, exist for converting solar energy into electricity, as shown in Fig. 13.1. The main emphasis of this chapter will be on three techniques that are

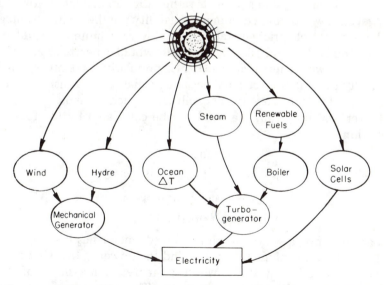

FIG. 13.1. METHODS FOR CONVERSION OF SOLAR ENERGY TO ELECTRICITY

not widely used at present, but which show some promise for future application as either supplemental or replacement methods. Use of plants or other biomass may be used as renewable fuel for electrical generation. However, this chapter will deal only with systems which generate electricity directly.

13.2 WIND ENERGY POTENTIAL

Unequal heating of the earth's surface by the sun produces air masses of differing temperature and density (reflected in the barometric pressure) and creates a simple atmospheric heat machine which produces air movements or wind. Tornados and hurricanes represent the extreme capabilities of sunlight to be transformed into wind forces. The problem is how to use wind since it is variable in time and intensity.

Probably the best known examples of the use of wind power are the Dutch windmills. Windmills were used extensively in Europe from the 12th through 18th centuries. Historically, pumping water has been the main function of windmills. They are still used in many remote range areas for stock watering. Pumping with windmills continues to be practical due to the relatively low power requirements. Multi-bladed units convert wind to rotary motion to drive a piston pump. Also, the water tank acts as a reservoir during calm periods. These pumping systems have a rather low efficiency, below 10%. However, low cost and equipment reliability are the main advantages.

Wind driven electric generating plants were common in rural areas during the 1930's and 1940's. Each unit would generate around 3000 watts of power. These plants lost their popularity as relatively cheap, centrally generated electricity became available to farmsteads and as farmstead loads outgrew the capabilities of such systems.

Power available from the wind can be calculated by the following equation:

$$P = 1.30 \times 20^{-2} A\, V^3$$

where
P = theoretical power (W)

A = area swept by the propeller (m^2)

V = wind velocity (km/h).

As an example, a windmill with a 3 m diameter has a sweep area of 7.1 m^2. In a 29 kilometers/hour wind the power in the wind for this area would be 735 watts. However, a theoretical maximum of 59.2% of this can be collected by a propeller. This would reduce the maximum collectible power to 435 watts. Since a well designed propeller

TABLE 13.1.   AVAILABLE POWER (WATTS) FROM WIND FOR SYSTEM WITH EFFI-
CIENCY OF 30%

Propeller Diameter	Wind Velocity (km/h)			
	10	20	30	40
1 m	3	24	83	196
2 m	12	98	331	784
3 m	28	221	744	1764
5 m	76	613	2068	4901
7 m	150	1200	4052	9606

may obtain approximately 75% of the maximum available power, the
power at the output shaft is reduced to 326 watts. This figure would
be further reduced in the conversion of the shaft power to electrical
power. An overall efficiency of approximately 30% could be expected from a wind generator system. Table 13.1 indicates the power
available from the wind assuming a system efficiency of 30%.

Another factor affecting total power output is the gustiness of the
winds. Due to gusts the energy actually generated over a period of
time may be twice that calculated from the average wind velocity.
This is due to the fact that power output is proportional to the cube
of the velocity, therefore, high velocity gusts have great power
potential.

Figure 13.2 indicates how the average available wind power varies

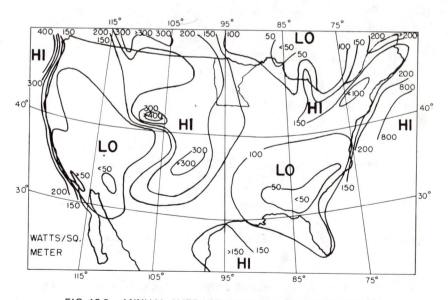

FIG. 13.2.   ANNUAL AVERAGE OF AVAILABLE WIND POWER

across the United States. It remains to be seen how much total available wind power can really be utilized. Much depends on the size and number of windmills and their location. One of the challenges of using wind energy is to design systems that work effectively in conjunction with existing power distribution systems.

13.3 SOLAR THERMAL CONVERSION

Lenses and mirrors can focus the sun's rays to create high enough temperatures to set materials on fire. Therefore, it seems logical that the same method could also be used to create steam. This is the basis for concentrating solar radiation by lenses, curved mirrors, or other collectors such that steam can be produced which will power a turbogenerator that produces electricity. Sunlight is a very diffuse energy source and must be collected over a wide area. It is estimated that a 1000 megawatt plant working at an average capacity of 60% in the southwestern part of the U.S. would need approximately 26 sq kilometers (10 sq miles). The initial cost of such a system likely would be several times that of a conventional power plant.

The principal aim of research and development in this area is to find ways to build such systems that will operate reliably at costs low enough to make them economically attractive.

Solar thermal conversion systems are made up of the following basic elements: 1) a concentrator to focus the sun's rays, 2) a receiver to absorb the solar energy, 3) a means for transmitting the heat to either a storage facility or directly to the turbogenerator, 4) a means of storing the heat for use at night or while the sun is not shining, and 5) a turbogenerator that converts the heat energy of the steam into electricity. Development of economically attractive methods of storage is probably the most limiting technical component at this time.

13.4 PHOTOVOLTAIC CONVERSION (SOLAR CELLS)

In certain substances, the absorption of light creates an electrical voltage that can be used as a power source. This process is called the photovoltaic effect. At the present time, silicon, cadmium sulfide, cadmium telluride, and gallium aresenide are the materials commonly used.

Solar cells offer a potentially attractive means for the direct conversion of sunlight into electricity with high reliability and low maintenance, as compared with solar-thermal systems. The present disadvantages are the high cost and the difficulty of storing large amounts of electrical energy for later use.

The most familiar applications of solar cells have been in the space program and in photographic light meters.

Solar cells also have important potential applications for power units where commercial power cannot be readily obtained. Power for remote sensing devices, harbor and buoy lights, fire telephones, and microwave repeater stations can be obtained through solar cells. In most applications, the cells will be connected to batteries which store energy for use when the sun is not shining.

To make large central installations of solar cells economically attractive, the cost must be reduced substantially. If the cost of solar cells is sufficiently reduced and economical storage techniques become available, solar cells could be used as elements in the central power system of a utility network or as on-site power supplies for individual residences or buildings.

Solar cells in use today have a conversion efficiency of approximately 10%. The maximum possible efficiency for conversion of radiant energy in sunshine to electricity by a solar cell is about 25%. To give an indication of the potential solar energy which can be calculated, assume a solar collector 6 m by 10 m is constructed. Using an average figure of 187 watts of solar energy reaching the ground per square meter and applying the 10% efficiency factor the average output of the collector would be 1122 watts.

REFERENCES

EATON, W. W. 1976. Solar Energy. Energy Research and Development Administration, Office of Public Affairs, Washington, D.C.

PRICE, E. R. and WEEKS, S. A. Wind Energy Systems. Agricultural Engineering Extension Bulletin 413. Cornell Univ., Ithaca, N.Y.

WILLIAMS, J. R. 1975. Solar Energy Technology and Applications. Ann Arbor Science Publications, Ann Arbor, Mich.

14

Introduction to Solid State Electronics

Solid state electronics are playing an increasingly important role in agriculture and in our general lives. Devices with solid state components are used for controls, communication equipment, indicators and in a wide variety of other applications. Solid state electronics have allowed the development of many useful tools for agriculture such as tractor, sprayer, combine and planter monitors, motor and lighting controls, calculators and computers.

The basic function of the solid state devices is to control or regulate the flow of electricity by the use of semiconductor materials. Development of semiconductors was initiated by the invention of the transistor in 1948. Starting with early application to radios and hearing aids, transistors have completely revolutionized the electronics industry. The transistor represented a major improvement in electronic components compared to the earlier vacuum tubes. Although they perform basically the same types of tasks, the transistor has many advantages which include: it requires no heater current, therefore power requirements are much smaller; it is very small and light, it is mechanically sturdy and long lasting; it can operate at low voltages yet carry relatively high currents; and it is thousands of times more reliable than the vacuum tube. Today the widespread availability of large and small computers is a result of the development of transistors and related solid-state devices such as diodes, power rectifiers, control rectifiers, photodevices, etc. The more recent development of integrated circuits, where complete electronic circuits are produced on minute pieces of semiconductor material, has lead to even wider application of semiconductors.

It is the intent of this chapter to develop a basic understanding of what semiconductors are and how they function. A few of the basic devices and sample applications will be discussed. More detailed description and theory of semiconductor circuits can be found in the references at the end of the chapter.

14.1 SEMICONDUCTOR STRUCTURE

In the study of semiconductors it is helpful to briefly review the atomic structure of matter. In a very simplified picture, an atom consists of a central nucleus containing positively charged protons with electrons in orbit around it. The negative charge of one electron is equal in magnitude to the positive charge of one proton. It is important to observe that all the electrons do not follow the same orbit. There is a set pattern of paths or rings located at different distances from the nucleus. Electrons located in rings close to the nucleus are tightly bound to it. However, the electrons in the outermost ring are comparatively loosely bound and can, under some situations, be dislodged. The electrical properties of the element are largely controlled by these outer ring or valence electrons.

If the number of electrons in the valence ring is less than four, the electrons are held to the atom rather loosely and can easily be moved from one atom to another. Materials made up of atoms with less than four electrons in their outer ring are termed conductors. Copper, with one electron in its outer ring, is an example of a good conductor.

If the number of electrons in the valence ring is greater than four, the electrons are held rather tightly and are not so easily dislodged. Such atoms make up materials termed insulators.

When the number of electrons in the valence ring is equal to four, the atom is neither a good conductor nor a good insulator and is termed a semiconductor. Two elements with four electrons in their

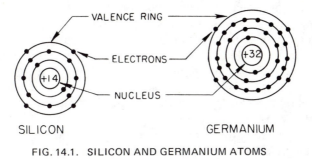

SILICON   GERMANIUM

FIG. 14.1.   SILICON AND GERMANIUM ATOMS

outer shell which are widely used for semiconductor materials are silicon and germanium.

Silicon atoms or germanium atoms will join with like atoms by means of covalent bonding to form pure solid crystals. Covalent bonding is the type of bonding in which the electrons in the outermost rings of the atoms are shared between atoms. For silicon and germanium the covalent bonding is such that each atom effectively has eight electrons in its outer ring making the outer ring complete. This makes the material a very good insulator since there appears to be more than four electrons in the outer ring.

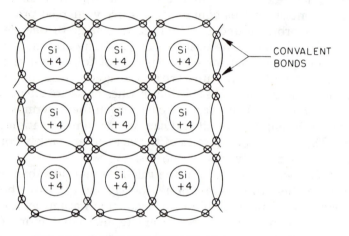

FIG. 14.2.   COVALENT BONDING OF SILICON ATOMS

Conductivity of the crystal can be changed in a controlled manner by adding specific impurities at the time of formation of the crystal. This process is called "doping". The new material is no longer a good insulator but possesses some very useful electrical properties.

One type of doping makes use of elements with five or more electrons in their valence ring such as phosphorus, arsenic, and antimony. When phosphorus, for example, is combined with silicon in covalent bonds there is one electron left over. This electron, called a "free" electron, can be made to move through the material very easily. This type of semiconductor material would be called a negative or N-type material.

The second type of doping makes use of elements with three or less valence electrons, such as boron and indium. When for example, boron is combined by covalent bonds with silicon there is a deficiency of one electron to complete the covalent bonding. This

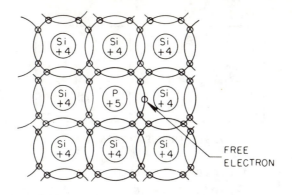

FIG. 14.3. AN N-TYPE SEMICONDUCTOR - SILICON DOPED
WITH PHOSPHORUS

electron deficiency is termed a "hole". Materials of this type are
termed positive or P-type materials.

In an N-type semiconductor, current flow is due to the movement
of free electrons. As a free electron moves away from the donor
atom where it originated it leaves behind a positive ion with a net
charge of +1. The positive ion is not free to move as it is fixed by the
crystal structure. Thus, an N-type material is dotted with fixed positive ions and mobile negative electrons. In the N-type semiconductor,
electrons are termed the majority carrier. In a similar manner a P-type
semiconductor may be considered to be dotted with fixed negative

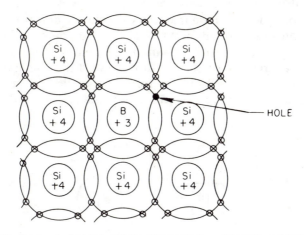

FIG. 14.4. A P-TYPE SEMICONDUCTOR - SILICON DOPED
WITH BORON

P – TYPE   N – TYPE

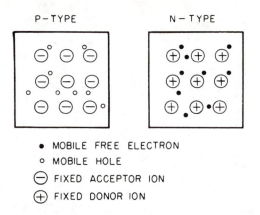

• MOBILE FREE ELECTRON
○ MOBILE HOLE
⊖ FIXED ACCEPTOR ION
⊕ FIXED DONOR ION

FIG. 14.5.   CURRENT CARRIERS IN SEMICONDUCTORS

ions and mobile positive holes. In a P-type semiconductor, holes are termed the majority carriers. In order to understand semiconductors it is necessary to accept the hole as a positively charged current carrier just as an electron is a negatively charged current carrier.

14.2 SEMICONDUCTOR DIODES

A semiconductor diode is a solid state device which will allow current to flow through itself in only one direction. A diode is formed when one part of a semiconductor is doped as an N-type material and an adjacent section is doped as a P-type material. The two parts are created in one piece of base material by a closely controlled diffusion process. The most common symbols for a diode are shown in Fig. 14.6. The arrowhead of the symbol designates the P-material and points in the direction of current flow, toward the N-material.

When the N and P sections are formed it would appear that the attraction between the free electrons and the holes would cause the

CURRENT FLOW

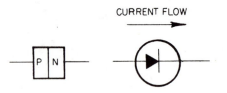

FIG. 14.6.   DIODE SYMBOLS

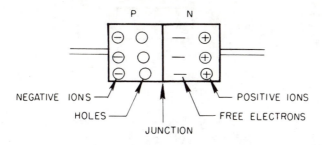

NEGATIVE IONS

HOLES

POSITIVE IONS

FREE ELECTRONS

JUNCTION

FIG. 14.7.   UNBIASED JUNCTION DIODE

electrons to drift across the junction and fill the holes. However, as some of the electrons drift across and combine with holes the positive ions exert an attractive force on the remaining electrons which holds them away from the junction. Similarly as holes move toward the junction, the negative ions exert an attraction on the remaining holes to prevent them from crossing the junction. The net result is a very rapid stability with a deficiency of electrons and holes at the junction.

Consider a circuit such as shown in Fig. 14.8, where the positive terminal of a battery is connected to the P-material of the diode and the negative terminal to the N-material. Since like charges will tend to repel each other, the battery tends to force the holes and free electrons toward the junction area of the diode. As electrons and holes congregate and combine at the junction, the battery will inject electrons in the N-material and attract them from the P-material to maintain a given rate of electron movement or current. The circuit arrangement illustrated with the battery negative terminal connected to the N-material and the battery positive to the P-material is known as the *forward bias* connection. The battery will cause holes and

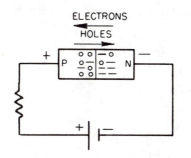

FIG. 14.8.   FORWARD BIASED DIODE

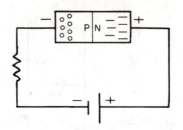

FIG. 14.9.   REVERSE BIASED DIODE

electrons to congregate at the junction which is the necessary condition for current to flow through the diode.

If the battery connections are reversed, that is the negative of the battery is connected to the P-material and the positive terminal to the N-material, the battery will tend to pull the holes and electrons away from the junction. Since in this situation no holes and electrons will be congregating or combining at the junction, no current will flow through the diode. This type of connection is called *reverse bias* and causes the diode to block current flow.

The characteristics of a diode can be summarized in a voltage vs. amperage characteristic curve as in Fig. 14.10. As the forward bias voltage increases the diode begins to conduct at a low voltage with

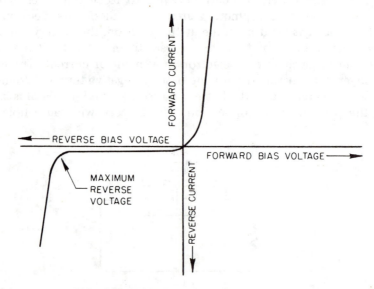

FIG. 14.10.   DIODE CHARACTERISTIC CURVE

rapid increases in current for small increases in voltage, that is, the forward bias resistance is low. For the reverse bias situation, the current flow remains small as the voltage level increases, that is, the reverse bias resistance is very high. Eventually the voltage will reach the maximum reverse voltage for the diode. Beyond this point the covalent bonds of the semiconductor material will begin to break down and a sharp rise in current will occur. If the reverse current is sufficient in magnitude and duration, the diode will be damaged due to excessive heat load. There is a type of diode, called the zener diode, which is designed to satisfactorily conduct current in the reverse direction under controlled conditions. This type of diode is used in some control circuits.

14.3 EXAMPLE DIODE APPLICATIONS

One of the most common uses of diodes is for rectification, that is, converting an AC power source to a DC power source. In the simplest case this can be done with one diode as shown in Fig. 14.11. Since the diode will only conduct when forward biased, current will flow through the load only during one-half of the AC cycle. This type of circuit is called a half-wave rectifier and produces a pulsating DC current through the load.

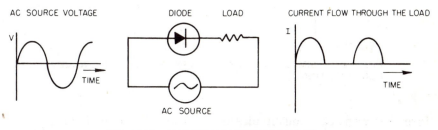

FIG. 14.11.   HALF-WAVE RECTIFICATION

Full-wave rectification can be accomplished by the use of four diodes connected in a diamond or bridge pattern. An example of this is the simple battery charger circuit shown in Fig. 14.12. Note that during each part of the AC cycle two of the diodes will be forward biased. Starting with the portion of the AC cycle where point A is positive relative to Point B, diodes D_1 and D_4 are forward biased and

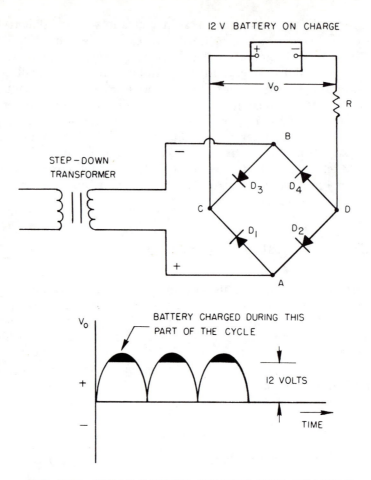

FIG. 14.12. SIMPLE BATTERY CHARGER WITH FULL-WAVE RECTIFICATION

form a current carrying circuit through the load while diodes D_2 and D_3 are reversed biased and do not conduct. When the orientation of A and B changes for the second half of the cycle, diodes D_2 and D_3 are forward biased and form the conducting circuit while D_1 and D_4 are reverse biased. The pattern of the diodes is such that the current flow through the battery is always in the same direction. This yields a full-wave pulsating DC signal going into the battery. It is also important to note that the battery is connected in such a manner that it cannot force current flow back through the circuit. The battery is reverse biasing the diodes at all times.

14.4 TRANSISTORS

A simple transistor is a solid state electronic device that is used in circuits to control current flow by the use of two PN junctions. The transistor is made up of three sections. Depending upon the manufacturing process they can be either NPN or PNP in nature. The term transistor comes from the combination of terms transfer and resistor which describes the device.

A transistor is generally shown diagrammatically in one of two ways. The upper diagram shown in Fig. 14.13 emphasizes that the transistor is made up of three sections termed the base, the emitter and the collector. The base is always the center section and is doped differently than the collector and emitter. The majority carriers in a PNP transistor are holes and those in an NPN transistor are free electrons. If a relatively small number of the majority carriers are withdrawn from the base region, current will flow through the transistor. As a small number of majority carriers are withdrawn from the base, the emitter "emits" a relatively large number of the majority carriers. The function of the collector is to "collect" the majority carriers not extracted from the base and then pass them on through the load circuit. This process will be further explained in the following examples. It should be noted that in an actual transistor the base region is very thin in comparison to the emitter and collector regions. It has been magnified in Fig. 14.13 only for illustration purposes.

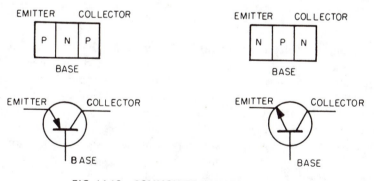

FIG. 14.13. COMMON TRANSISTOR SYMBOLS

For the type of symbol shown in the lower portion of Fig. 14.13, the arrowhead is always placed on the emitter. The arrowhead points towards the N-type material and away from the P-type material (in the direction of current flow). The base connection is illustrated as a center lead which comes out at a right angle. The collector lead is the lead opposite the emitter lead. It can be determined if the transistor is PNP or NPN by following the simple rule "if the arrow points toward the base, the transistor is PNP; if the arrow points away from the base, the transistor is NPN."

With a battery connected to a PNP transistor in the orientation shown in Fig. 14.14, switch S1 closed and switch S2 open, current will flow through the emitter and base of the transistor. With the S2 switch open, the collector is inoperative and the transistor acts as a simple PN junction diode (the emitter and base) connected to the battery in a forward bias orientation. All current flow is through switch S1.

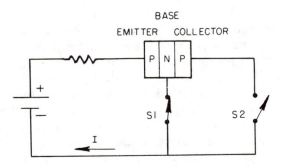

FIG. 14.14.   TRANSISTOR USED AS A SIMPLE DIODE

Note that to fully understand the operation of a PNP transistor it is necessary to accept hole movement as a method of current flow and apply it to the transistor. Using the hole movement concept, the current flow in the transistor is a result of the movement of holes through the P material to the N material.

When the switch S2 is closed as shown in Fig. 14.15, the total current remains unchanged; however, now most of the current (approximately 95%) leaves the transistor through the collector. This means the base current is reduced to a very small value.

When the switches are in the positions shown in Fig. 14.16, where S2 is closed and S1 is open, no appreciable current will flow in the circuit. With the base circuit open there are no holes being injected into the base from the emitter and consequently there are no holes in the base which can be attracted by the negative battery potential

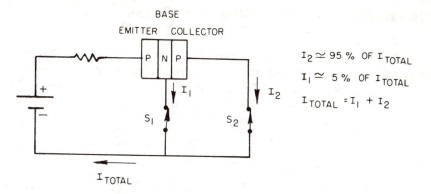

FIG. 14.15.   TRANSISTOR IN NORMAL OPERATING MODE

into the collector. When no current is flowing through the base connected circuit, the resistance between the emitter and collector appears to be very high. Opening switch S1 effectively "shuts off" the
transistor so that no appreciable current will flow.

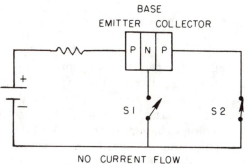

NO CURRENT FLOW

FIG. 14.16.   TRANSISTOR IN THE OFF POSITION

Note that an NPN transistor operates in the same manner as a PNP,
except the current consists of the movement of electrons from the
emitter to the base and collector, and voltage polarity of the source
is reversed.

Transistors act as current regulators in an electrical circuit. In ac-
complishing this function, they can act as either an amplifier or as a
switch. The following examples will make use of our understanding
of transistors and demonstrate both types of uses.

14.5 EXAMPLE TRANSISTOR APPLICATIONS

A circuit showing how a transistor can be used as part of the voltage regulating system for an automotive type generator is shown in Fig. 14.17. The purpose of the regulator is to intermittently open the circuit to the generator field winding in order to limit the voltage level output of the generator to a preset value.

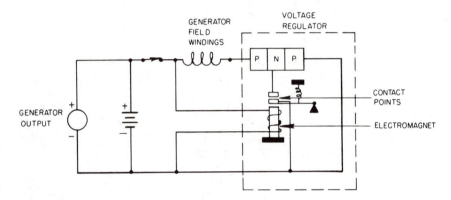

FIG. 14.17.   GENERATOR CIRCUIT WITH TRANSISTORIZED VOLTAGE REGULATOR

The output voltage of the generator is controlled by the strength of the field produced by the generator field windings. As the current in the field windings and the field strength increases, the voltage level of the generator output increases. For the system shown in Fig. 14.17 as the generator voltage increases the pull of the electromagnet increases until it causes the contacts in the base connected circuit of the transistor to open. With the contacts open there is no transistor base current therefore no collector current will flow. The transistor "shuts off" the generator field current and the generator voltage will decrease. After the voltage has dropped sufficiently the contacts will reclose and current again flows in the field winding. The cycle of opening and closing the contacts occurs many times per second.

In a nontransistorized regulator, full generator field current runs through the controlled set of contacts. In the transistorized regulator the amount of current through the contact points is very small. Since the amount of current which can be controlled is limited by the contact current, the transistorized system allows the use of larger field currents and improves the life of the contacts.

Figure 14.18 demonstrates the use of a transistor as an amplifier in a microphone or loudspeaker circuit. In this example, electrical con-

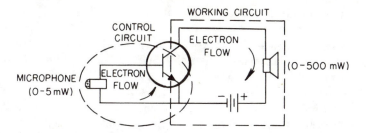

FIG. 14.18. AMPLIFIER CIRCUIT USING A TRANSISTOR

trol comes from the microphone whose fluctuating electric current output corresponds to the fluctuations in sound waves hitting the microphone. However, a microphone can produce only a small amount of power, therefore an amplifier is needed to raise the level of the signal before it is input to a speaker.

Assume for purposes of illustration, the microphone power output is in the range of 0 to 5 milliwatts (mW) and the power produced by the battery in the main circuit is capable of varying from 0 to 500 milliwatts. If a sound wave hits the microphone and creates a power output of 3 milliwatts, the microphone allows a flow of holes from the emitter to the base. The result of the base current is a relatively large current flow from the emitter to the collector and through the loudspeaker. In this manner the current through the loudspeaker is controlled in exact proportion to the much smaller microphone signal. If the collector current is 100 times the base current, the transistor is able to amplify the signal 100 times. For the example, the power to the speaker would be 300 milliwatts for a microphone current of 3 milliwatts.

The proportion between the power before amplification and that afterwards is reasonably constant in most transistorized amplifiers, which allows the amplifier to work for a broad range of input signals.

14.6 THYRISTORS

A number of other devices have been developed as outgrowths of the basic diode and transistor. Thyristors are semiconductor switching devices that do not require any control or base-type current once they are turned on. Two of the most important types of thyristers are silicon controlled rectifiers (SCRs) and TRIACs. These types of thyristors are generally capable of controlling relatively large amounts of power without reaching harmful temperatures. They are therefore commonly found in circuits for control of motors and other loads acting as solid state relays.

An SCR, as shown in Fig. 14.19, normally blocks current flow in both directions. However when the SCR is forward biased, a quick pulse of current into the gate will turn on the SCR. The SCR keeps conducting or stays on, even without a gate current, as long as the device is forward biased. When the device turns off, because it is no longer forward biased, it requires another gate signal to start conducting again.

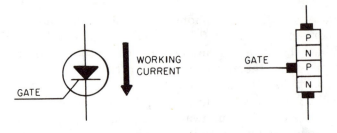

FIG. 14.19. SILICON CONTROLLED RECTIFIER (SCR)

Figure 14.20 shows conceptually how an SCR may be used in a solid-state auto ignition system. The DC power supply system stores up energy in the capacitor while the SCR is turned off. When the breaker points momentarily close, a small current is sent into the gate of the SCR. This turns on the SCR and allows the charge of the capacitor to force a high surge of current through the spark coil. The spark coil acts as a step-up transformer. The high voltage poten-

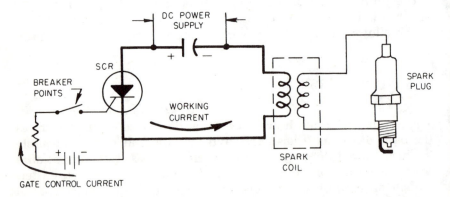

FIG. 14.20. SIMPLIFIED SOLID STATE IGNITION SYSTEM

tial across the plug then causes a spark discharge across the plug gap. The capacitor is quickly discharged and the current ceases to flow, therefore the SCR turns off and the capacitor is recharged.

The chief advantage of the SCR in this application is that the breaker points carry only a very small current at a low voltage. Therefore, compared to conventional ignition systems where the points control the whole working current, the points almost never wear out. In addition, higher voltages and currents can be used in the load portion of the circuit giving a hotter spark.

The TRIAC is much like a pair of SCRs connected head-to-toe as shown in Fig. 14.21. Note that the gate terminal is common to both SCR's. Normally the TRIAC will block current flow in both directions. However it can be triggered to conduct in either direction by a momentary pulse applied to the gate. This characteristic is reflected in its formal name "bidirectional triode thyristor".

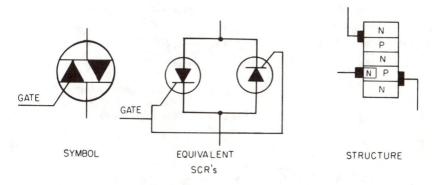

FIG. 14.21. BIDIRECTIONAL TRIODE THYRISTOR (TRIAC)

As shown in Fig. 14.21, the TRIAC has five basic layers and a small sixth region under a portion of the gate contact. This gate arrangement allows the TRIAC to be triggered by gate currents in either direction in contrast to the SCR where current must flow into the gate region.

One common application of the TRIAC is in speed control for AC motors. As shown in Fig. 14.22, an adjustable triggering circuit is established which triggers the TRIAC by sending current through the gate circuit sometime within each half-cycle of the AC source. The earlier in the half-cycle the TRIAC is triggered the more current that is allowed to pass through the motor and the faster it runs. This type of control is called "phase-control".

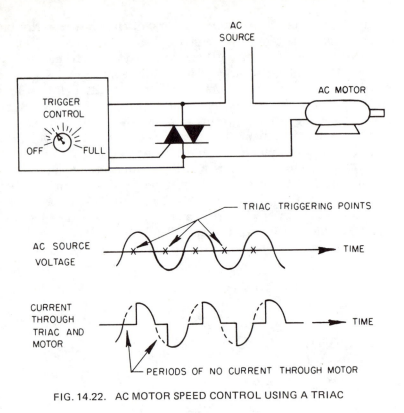

FIG. 14.22.   AC MOTOR SPEED CONTROL USING A TRIAC

14.7 INTEGRATED CIRCUITS

So far this chapter has discussed what are termed discrete semiconductor devices. Each diode or transistor is a discrete device connected in a circuit by appropriate electrical conductors. Another advancement in electronics has been the development of integrated circuits. An integrated circuit (IC) is a complete electronic circuit contained entirely within a single chip of silicon. The IC may contain transistors and perhaps diodes, resistors and capacitors along with their interconnecting electrical conductors. The major advantages of having an electronic circuit in integrated form rather than made up of discrete components are smaller size, lower cost, and higher reliability.

Small scale integrated circuits (SSI), medium scale integrated circuits (MSI), and large scale integrated circuits (LSI) are some of the classifications of circuits being developed. The present era in elec-

tronics is being dominated by integrated circuits, particularly the LSI
type where several thousand components may be placed on one chip
with a volume of about 2×10^{-3} cm^3.

Most IC's are formed by starting with a silicon chip doped as a
P-type semiconductor. The desired elements and connections are
then formed in the chip by precisely controlled diffusion processes.
The basic form of an IC type resistor, a capacitor and a simple transistor are shown in Fig. 14.23.

(a)   INTEGRATED RESISTOR

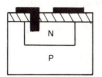

METAL LEADS

INSULATING LAYER

RESISTANCE IS DETERMINED BY SHAPE,
SIZE, AND CONDUCTIVITY OF MATERIAL
BETWEEN THE TWO LEADS.

(b)   INTEGRATED CAPACITOR

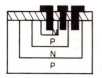

CAPACITANCE DEVELOPED BY USE OF
INSULATING OXIDE LAYER AS
DIELECTRIC BETWEEN THE TWO
LEADS (PLATES).

(c)   INTEGRATED NPN TRANSISTOR

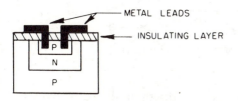

FIG. 14.23. BASIC CIRCUIT ELEMENTS FOR INTEGRATED CIRCUITS

A simple integrated circuit containing a resistor, a diode, and an
NPN transistor is shown in cross section and diagrammatically in Fig.
14.24. The gold or aluminum connectors on the surface interconnect
the components in the desired sequence. Fine gold wires would be
connected to the metal film at the appropriate locations to give the
external connections.

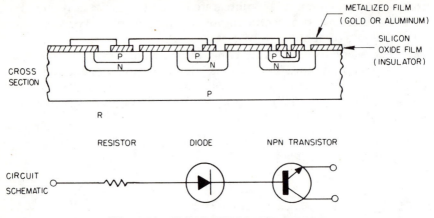

FIG. 14.24. SIMPLE INTEGRATED CIRCUIT

Since all the components are formed on the same slice of silicon, it might appear that they could short-circuit through the silicon. However the cross section view shows how each component is isolated (insulated) from the others by action of PN junctions. All possible short-circuit current paths are blocked by combinations of PN junctions. The result is that the current is forced to take the intended path through the components.

REFERENCES

ANON. 1968. Fundamentals of Semiconductors. Training Chart Manual Section J., Delco Remy, Anderson, IN.

ANON. 1978. Understanding Solid-State Electronics. 3rd Edition. Texas Instruments Learning Center. Dallas, Tex.

BUTCHBAKER, A. F. 1976. Electricity and Electronics for Agriculture. Iowa St. Univ. Press, Ames, Ia.

HIBBERD, R. G. 1968. Solid-State Electronics. McGraw-Hill, N.Y.

Answers to Selected Problems

CHAPTER 1

2. nine
3. 57.6 ohms
4. 40 watt bulb
5. $0.36
6. 8.33 amperes
7. (a) 500 V   (b) 232.5   (c) 17.2 V

CHAPTER 2

2. (a) I_1 = 5 amp I_2 = 2 amp I_3 = 3.33 amp
   (b) E_1 = 20 V E_2 = 50 V E_3 = 30 V
3. (a)

Resistor	Voltage Drop	Current Flow
1	90 V	30 amp
2	30	10
3	30	5
4	30	10
5	10	5
6	20	5

(b)

Resistor	Voltage Drop	Current Flow
1	60 V	3.0 amp
2	15	1.5
3	15	1.0
4	15	0.5
5	30	3.0

(c) Resistor	Voltage Drop	Current Flow	(d) Resistor	Voltage Drop	Current Flow
1	25 V	10 amp	1	90 V	3.0 amp
2	20	6.33	2	30	2.0
3	20	3.66	3	30	1.0
4	30	15			
5	15	7.5			
6	15	7.5			

4. 9,970 ohms

5. (a) & (b)

Bulb	Resistance	Current
75 W	192 ohms	0.625 amp
100	144	0.833
150	96	1.25
300	48	2.50

   (c)   5.05 kW
   (d)   42.1 amp
   (e)   (Recall 1W = 1 J/s)   3,535 J/s

CHAPTER 3

1. (a) $Z = 32.3$ ohms
   (b) $I = I_R = I_L = I_C = 3.72$ amp
   (c) $E_R = 63.2$ V $E_C = 31.0$ V $E_L = 133.0$ V
2. (a) $E_S = 35$ V   (d) $Z = 3.5$ ohms
   (b) $R = 2.1$ ohms   (e)   p.f. = 0.6
   (c) $X_L = 2.8$ ohms   (f)   T.P. = 210 W
3. (a) $I_R = 12$ amp $I_C = 12$ amp $I_L = 20$ amp
   (b) $I_T = 13.4$ amp
   (c)   p.f. = 0.896
   (d)   T.P. = 1441 W
   (e) $Z = 8.96$ ohms
4. (a) $C = 922.6 \, \mu f$
   (b) $C = 587.5 \, \mu f$
5. 20.0 kVar of capacitance

CHAPTER 4

1.

| | Current in Line (amps) | | |
Load Connected	Hot 1	Neutral	Hot 2
A	10	10	0
A&D	10	0	10
A&C	10	5	15
A, C&E	32	5	37
A, B, C, &D	26	1	25
All	98	1	97

2. (a) I_L = 43.3 amp E_L = 240 V
   (b) 16,200 W
3. (a) I_L = 25 amps E_L = 415.7 V
   (b) 16,200 W

CHAPTER 8

1. (a) p.f. = 0.78
   (b) T.P. = 2238 W = 3.0 hp
   (c) Power Out = 1678 W = 2.25 hp
2. (a) T.P. = 3264 W (b) EFF = 68.6% (c) $9.79
3. 1000 RPM
4. (a) 39,750,000 ft-lb (b) 20 hp (c) 44.4 hp
5. No. 10 THW Copper
6. No. 8 UF Aluminum

CHAPTER 9

1. Initial = 14,400 lm 70% of Rated Life = 12,816 lm
2. 3 lamps
7. 10 lx
8. 3.9 m

Appendix A

TABLE A.1.   PROPERTIES OF ANNEALED COPPER WIRE

Size B & S[1] or AWG[2]	Diameter		Lb per 1000 ft	Kg per 1000 m	Ohms per 1000 ft @ 20°C	Ohms per 1000 m @ 20°C
	mils	cm				
0000	460	1.17	641	954	0.049	0.161
000	410	1.04	508	756	0.0618	0.202
00	365	0.927	403	600	0.0779	0.256
0	325	0.894	320	476	0.0983	0.322
1	289	0.734	253	376	0.124	0.406
2	258	0.655	201	299	0.156	0.513
4	204	0.518	126	188	0.248	0.815
6	162	0.411	79.5	118	0.395	1.30
8	128.5	0.326	50.0	74.4	0.628	2.06
10	101.9	0.259	31.4	46.7	1.000	3.28
12	80.8	0.205	19.8	29.5	1.588	5.21
14	64.1	0.163	12.4	18.4	2.525	8.28
16	50.8	0.129	7.82	11.6	4.02	13.18

[1] B & S — Brown and sharpe gauge
[2] AWG — American wire gauge

TABLE A.2.   PROPERTIES OF ALUMINUM WIRE

Size B & S[1] or AWG[2]	Diameter		Lb per 1000 ft	Kg per 1000 m	Ohms per 1000 ft @ 20°C	Ohms per 1000 m @ 20°C
	mils	cm				
0000	528	1.34	199	296	0.0809	0.265
000	470	1.20	157	234	0.102	0.335
00	418	1.06	125	186	0.129	0.423
0	372	0.945	99.1	147	0.162	0.531
1	332	0.843	78.6	117	0.204	0.669
2	292	0.742	62.3	92.7	0.258	0.846
6	184	0.467	24.2	36.0	0.640	2.10
8	128	0.325	15.2	22.6	1.02	3.35
10	102	0.259	9.56	14.2	1.62	5.31
12	80.8	0.205	6.01	8.94	2.57	8.43

[1] B & S — Brown and Sharpe Gauge
[2] AWG — American Wire Gauge

TABLE A.3. FULL-LOAD CURRENTS FOR SINGLE-PHASE MOTORS

HP	Single-Phase AC Motors 115 V	230 V
1/6	4.4 amp	2.2 amp
1/4	5.8	2.9
1/3	7.2	3.6
1/2	9.8	4.9
3/4	13.8	6.9
1	16	8
1 1/2	20	10
2	24	12
3	34	17
5	56	28
7 1/2	80	40
10	100	50

Source: Anon. (1977).
Note: The voltages are rated motor voltages. The currents listed shall be permitted for system voltage ranges of 110 to 120 and 220 to 240.

TABLE A.4. FULL-LOAD CURRENT FOR THREE-PHASE MOTORS

HP	Three-Phase AC Motors Induction Type Squirrel-Cage and Wound Rotor 115 V	230 V	460 V
1/2	4 amp	2 amp	1 amp
3/4	5.6	2.8	1.4
1	7.2	3.6	1.8
1 1/2	10.4	5.2	2.6
2	13.6	6.8	3.4
3		9.6	4.8
5		15.2	7.6
7 1/2		22	11
10		28	14
15		42	21
20		54	27
25		68	34

Source: Anon. (1977).

TABLE A.5. ALLOWABLE AMPACITIES OF INSULATED COPPER CONDUCTORS

Single Conductor in Free Air (Based on Ambient Temperature of 30°C [86°F])

Size AWG MCM	60°C (140°F) Types RUW (14-2), T, TW	75°C (167°F) Types RH, RHW, RHW (14-2), THW, XHHW	85°C (185°F) Types V, MI	90°C (194°) Types TA, TBS, SA, AVB, SIS, FEP, FEPB, RHH, THHN, XHHW	Bare and Covered Conductors
18	—	—		25	—
16	—	—	27	27	—
14	20	20	30	30	30
12	25	25	40	40	40
10	40	40	55	55	55
8	55	65	70	70	70
6	80	95	100	100	100
4	105	125	135	135	130
3	120	145	155	155	150
2	140	170	180	180	175
1	165	195	210	210	205
1/0	195	230	245	245	235
2/0	225	265	285	285	275
3/0	260	310	330	330	320
4/0	300	360	385	385	370
250	340	405	425	425	410
300	375	445	480	480	460
350	420	505	530	530	510
400	455	545	575	575	555
500	515	620	660	660	630

Not More Than Three Conductors in Raceway or Cable or Direct Burial (Based on Ambient Temperature of 30°C [86°F])

Size AWG MCM	60°C (140°F) Types RUW (14-2), T, TW, UF	75°C (167°F) Types RH, RHW, RUH (14-2), THW, THWN, XHHW, USE	85°C (185°F) Types V, MI	90°C (194°F) Types TA, TBS, SA, AVB, SIS, FEP, FEPB, RHH, THHN, XHHW
18	—	—	—	21
16	—	—	22	22
14	15	15	25	25
12	20	20	30	30
10	30	30	40	40
8	40	45	50	50
6	55	65	70	70
4	70	85	90	90
3	80	100	105	105
2	95	115	120	120
1	110	130	140	140
1/0	125	150	155	155
2/0	145	175	185	185
3/0	165	200	210	210
4/0	195	230	235	235
250	215	255	270	270
300	240	285	285	300
350	260	310	325	325
400	280	335	360	360
500	320	380	405	405

Source: Anon. (1977).

TABLE A.6. ALLOWABLE AMPACITIES OF INSULATED ALUMINUM AND COPPER-CLAD ALUMINUM CONDUCTORS

Single Conductor in Free Air (Based on Ambient Temperature of 30°C [86°F])

Size AWG MCM	Temperature Rating of Conductor				
	60°C (140°F) Types RUW (12-2), T, TW	75°C (167°F) Types RH, RHW, RUH (12-2), THW, THWN, XHHW	85°C (185°F) Types V, MI	90°C (194°) Types TA, TBS, SA, AVB, SIS, RHH, THHN, XHHW*	Bare and Covered Conductors
12	20	20	30	30	30
10	30	30	45	45	45
8	45	55	55	55	55
6	60	75	80	80	80
4	80	100	105	105	100
3	95	115	120	120	115
2	110	135	140	140	135
1	130	155	165	165	160
1/0	150	180	190	190	185
2/0	175	210	220	220	215
3/0	200	240	255	255	250
4/0	230	280	300	300	290
250	265	315	330	330	320
300	290	350	375	375	360
350	330	395	415	415	400
400	355	425	450	450	435
500	405	485	515	515	490

Not More Than Three Conductors in Raceway or Cable or Direct Burial (Based on Ambient Temperature of 30°C [86°F])

Size AWG MCM	Temperature Rating of Conductor			
	60°C (140°F) Types RUW (12-2), T, TW, UF	75°C (167°F) Types RH, RHW, RUH (12-2), THW, THWN, XHHW, USE	85°C (185°F) Types V, MI	90°C (194°F) Types TA, TBS, SA, AV, SIS, RHH, THHN, XHHW*
12	15	15	25	25
10	25	25	30	30
8	30	40	40	40
6	40	50	55	55
4	55	65	70	70
3	65	75	80	80
2	75	90	95	95
1	85	100	110	110
1/0	100	120	125	125
2/0	115	135	145	145
3/0	130	155	165	165
4/0	155	180	185	185
250	170	205	215	215
300	190	230	240	240
350	210	250	260	260
400	225	270	290	290
500	250	310	330	330

Source: Anon. (1977).

TABLE A.7. CONDUCTOR INSULATION DEFINITIONS AND APPLICATIONS

Trade Name	Type Letter	Max. Operating Temp.	Application Provisions	Insulation
Heat-Resistant Rubber	RH	75°C 167°F	Dry locations.	Heat-Resistant Rubber
Heat-Resistant Rubber	RRH	90°C 194°F	Dry locations.	
Moisture and Heat-Resistant Rubber	RHW	75°C 167°F	Dry and wet locations. For over 2000 volts insulation shall be ozone-resistant.	Moisture and Heat-Resistant Rubber
Moisture-Resistant Latex Rubber	RUW	60°C 140°F	Dry and wet locations.	90% Unmilled, Grainless Rubber
Thermoplastic	T	60°C 140°F	Dry locations.	Flame-Retardant, Thermoplastic Compound
Moisture-Resistant Thermoplastic	TW	60°C 140°F	Dry and wet locations.	Flame-Retardant, Moisture-Resistant Thermoplastic
Heat-Resistant Thermoplastic	THHN	90°C 194°F	Dry locations.	Flame-Retardant, Heat-Resistant Thermoplastic
Moisture- and Heat-Resistant Thermoplastic	THW	75°C 167°F 90°C 194°F	Dry and wet locations. Special applications *within* electric discharge lighting equipment.	Flame-Retardant, Moisture- and Heat-Resistant Thermoplastic
Moisture- and Heat-Resistant Thermoplastic	THWN	75°C 167°F	Dry and wet locations.	Flame-Retardant, Moisture- and Heat-Resistant Thermoplastic
Moisture- and Heat-Resistant Cross-Linked Synthetic Polymer	XHHW	90°C 194°F 75°C 167°F	Dry locations. Wet locations.	Flame-Retardant Cross-Linked Synthetic Polymer

TABLE A.7. (Continued)

Trade Name	Type Letter	Max. Operating Temp.	Application Provisions	Insulation
Underground Feeder & Branch-Circuit Cable-Single Conductor	UF	60°C	Underground Feeder and Branch-Circuit Cable. (See also NEC Section 339)	Moisture-Resistant
		75°C 167°F		Moisture- and Heat-Resistant
Underground Service Entrance Cable-Single Conductor	USE	75°C 167°F	Underground Feeder and Branch-Circuit Cable. (See also NEC Section 338)	Heat- and Moisture-Resistant

Source: Anon. (1977).

TABLE A.8. POWER FACTOR IMPROVEMENT TABLE[1]

	KW Multipliers For Determining Capacitor KVAR's					
Original Power Factor	Desired Improved Power Factor					
	80%	85%	90%	92%	95%	100%
50	0.982	1.112	1.248	1.306	1.403	1.732
54	0.809	0.939	1.075	1.133	1.230	1.559
58	0.655	0.785	0.921	0.979	1.076	1.405
60	0.583	0.713	0.849	0.907	1.004	1.333
62	0.516	0.646	0.782	0.840	0.937	1.266
64	0.451	0.581	0.717	0.775	0.872	1.201
66	0.388	0.518	0.654	0.712	0.809	1.138
68	0.328	0.458	0.594	0.652	0.749	1.078
70	0.270	0.400	0.536	0.594	0.691	1.020
72	0.214	0.344	0.480	0.538	0.635	0.964
74	0.159	0.289	0.425	0.483	0.580	0.909
76	0.105	0.235	0.371	0.429	0.526	0.855
78	0.052	0.182	0.318	0.376	0.473	0.802
80		0.130	0.266	0.324	0.421	0.750
82		0.078	0.214	0.272	0.369	0.698
84		0.026	0.162	0.220	0.317	0.646
86			0.109	0.167	0.264	0.593
88			0.056	0.114	0.211	0.540
90				0.058	0.155	0.484
92					0.097	0.426
94					0.034	0.363
96						0.292
98						0.203
99						0.143

[1] Figures below X kilowatt input = KVar of capacitors required to improve from one power factor to another.

TABLE A.9.   TRIGONOMETRIC FUNCTIONS

Values of the Trigonometric Functions of Angles 0-90°

Angle	Sin	Cos	Tan	Angle	Sin	Cos	Tan
0°	0.00000	1.0000	0.00000	46°	0.71934	0.69466	1.0355
1°	0.01745	0.99985	0.01746	47°	0.73135	0.68200	1.0724
2°	0.03490	0.99939	0.03492	48°	0.74314	0.66913	1.1106
3°	0.05234	0.99863	0.05241	49°	0.75471	0.65606	1.1504
4°	0.06976	0.99756	0.06993	50°	0.76604	0.64279	1.1918
5°	0.08716	0.99619	0.08749	51°	0.77715	0.62932	1.2349
6°	0.10453	0.99452	0.10510	52°	0.78801	0.61566	1.2799
7°	0.12187	0.99255	0.12278	53°	0.79864	0.60182	1.3270
8°	0.13917	0.99027	0.14054	54°	0.80902	0.58779	1.3764
9°	0.15643	0.98769	0.15838	55°	0.81915	0.57358	1.4281
10°	0.17365	0.98481	0.17633	56°	0.82904	0.55919	1.4826
11°	0.19081	0.98163	0.19438	57°	0.83867	0.54464	1.5399
12°	0.20791	0.97815	0.21256	58°	0.84805	0.52992	1.6003
13°	0.22495	0.97437	0.23087	59°	0.85717	0.51504	1.6643
14°	0.24192	0.97030	0.24933	60°	0.86603	0.50000	1.7321
15°	0.25882	0.96593	0.26795	61°	0.87462	0.48481	1.8040
16°	0.27564	0.96126	0.28675	62°	0.88295	0.46947	1.8807
17°	0.29237	0.95630	0.30573	63°	0.89101	0.45399	1.9626
18°	0.30902	0.95106	0.32492	64°	0.89879	0.43837	2.0503
19°	0.32557	0.94552	0.34433	65°	0.90631	0.42262	2.1445
20°	0.34202	0.93969	0.36397	66°	0.91355	0.40674	2.2460
21°	0.35837	0.93358	0.38386	67°	0.92050	0.39073	2.3559
22°	0.37461	0.92718	0.40403	68°	0.92718	0.37461	2.4751
23°	0.39073	0.92050	0.42447	69°	0.93358	0.35837	2.6051
24°	0.40674	0.91355	0.44523	70°	0.93969	0.34202	2.7475
25°	0.42262	0.90631	0.46631	71°	0.94552	0.32557	2.9042
26°	0.43837	0.89879	0.48773	72°	0.95106	0.30902	3.0777
27°	0.45399	0.89101	0.50953	73°	0.95630	0.29237	3.2709
28°	0.46947	0.88295	0.53171	74°	0.96126	0.27564	3.4874
29°	0.48481	0.87462	0.55431	75°	0.96593	0.25882	3.7321
30°	0.50000	0.86603	0.57735	76°	0.97030	0.24192	4.0108
31°	0.51504	0.85717	0.60086	77°	0.97437	0.22495	4.3315
32°	0.52992	0.84805	0.62487	78°	0.98163	0.20791	4.7046
33°	0.54464	0.83867	0.64941	79°	0.98163	0.19081	5.1446
34°	0.55919	0.82904	0.67451	80°	0.98481	0.17365	5.6713
35°	0.57358	0.81915	0.70021	81°	0.98769	0.15643	6.3138
36°	0.58779	0.80902	0.72654	82°	0.99027	0.13917	7.1154
37°	0.60182	0.79864	0.75355	83°	0.99255	0.12187	8.1443
38°	0.61566	0.78801	0.78129	84°	0.99452	0.10453	9.5144
39°	0.62932	0.77715	0.80978	85°	0.99619	0.08716	11.430
40°	0.64279	0.76604	0.83910	86°	0.99756	0.06976	14.301
41°	0.65606	0.75471	0.86929	87°	0.99863	0.05234	19.081
42°	0.66913	0.74314	0.90040	88°	0.99939	0.03490	28.636
43°	0.68200	0.73135	0.93252	89°	0.99985	0.01745	57.290
44°	0.69466	0.71934	0.96569	90°	1.0000	0.0000	∞
45°	0.70711	0.70711	1.0000				

TABLE A.10. STANDARD INTERNATIONAL (SI) UNIT PREFIXES

Multiples & Submultiples	Prefix	SI Symbol
10^6	mega	M
10^3	kilo	k
10^2	hecto	h
10^1	deka	da
10^{-1}	deci	d
10^{-2}	centi	c
10^{-3}	milli	m
10^{-6}	micro	μ
10^{-9}	nano	n
10^{-12}	pico	p

TABLE A.11. STANDARD INTERNATIONAL (SI) – ENGLISH UNIT CONVERSION FACTOR TABLE

To Convert From	To	Multiply By
LENGTH – meter (m)		
foot (ft)	m	0.30480^1
inch (in.)	m	0.02540^1
mile (U.S. statute)	m	1,609.3
rod	m	5.02920^1
yard	m	0.91440^1
AREA – square meter (m^2)		
acre	m^2	4046.86
square foot	m^2	0.09290
square mile	km^2	2.58999
acre	hectare	0.40468
VOLUME – cubic meter (m^3)		
bushel (U.S.)	m^3	0.03524
cubic foot	m^3	0.02831
gallon	m^3	0.00378
cubic yard	m^3	0.76455
quart (U.S. liquid)	liter	0.94635
gallon (U.S. liquid)	liter	3.78541
FORCE – newton (N)		
pound-force (lbf)	N	4.44822
poundal	N	0.13825
dyne	N	0.00001^1
MASS – kilogram (kg)		
pound-mass (lbm)	kg	0.45359
ton (short, 2000 lbm)	kg	907.185
metric ton (t)	kg	1000.00^1
ENERGY (work) – joule (J)		
British thermal unit (Btu)	J	105.505
calorie (cal)	J	4.18680
kilowatt-hour (kWh)	J	$3,600,000.^1$
watt-second (Ws)	J	1.000000^1
foot-pound-force (ftlbf)	J	1.35581

TABLE A.11. (Continued)

To Convert From	To	Multiply By
POWER – watt (W)		
British thermal unit per hour	W	0.29307
foot-pound-force per second	W	1.35581
horsepower (550 ft·lbf/s	W	745.700
HEAT		
British thermal unit per square foot	J/m^2	11,356
British thermal unit per pound-mass	J/kg	$2{,}326.0^1$
LIGHT		
foot candle	lm/m^2	10.7639
lux	lm/m^2	1.000000^1
PRESSURE OR STRESS – pascal (Pa)		
atmosphere	Pa	101325.
inch of mercury (60°F)	Pa	3376.85
inch of water (60°F)	Pa	248.840
pound-force per square foot	Pa	47.8803
pound-force per square inch	Pa	6894.76

[1] Relationships are exact conversions

Appendix B

RESIDENTIAL ELECTRIC OUTLET PLANNING GUIDE[1]

This planning guide is a *room by room list of recommendations* to ensure adequate outlets in a residence. *It goes beyond the minimum requirements of the National Electrical Code.* This guide is intended to be used to plan the outlets for a residence and does not discuss methods or procedures for doing the wiring. Complete details are not given in some recommendations; however, appropriate code article and paragraph are listed.

On a number of occasions in this guide, National Electrical Code Article 210-25b is referred to as "the 2 meter (6 ft) rule." This rule for spacing receptacle outlets is as follows: Receptacle outlets shall be installed so that no point along the floor line in any usable wall space (a space 0.6 m (2 ft) wide or greater) is more than 2 meters from an outlet in that space. This rule gives the minimum number of receptacle outlets.

Living Room

Lighting. — General illumination — ceiling or wall fixtures, cove, valance or cornice lighting. Wall-switch (multiple) controlled. If none of the above, wire at least one receptacle outlet so it is controlled by multiple switches.

---

[1] Revised from Surbrook. (1973).

Receptacle Outlets. — Use the 2 meter rule.

Special Purpose Outlets. — Air conditioner

Dining Room

Lighting. — At least one wall-switch controlled lighting outlet, usually over probable location of table. Dimmer switch may be desired.

Receptacle Outlets. — Use the 2 meter rule. If table is to be placed against the wall, one outlet just above table. If built-in open counter space, an outlet above counter height for use of portable appliances. *These outlets must be on a 20 ampere circuit.*

Kitchen

Lighting. — General illumination, wall-switch controlled, lighting at sink, wall-switch controlled. Also, provide for illumination of work areas, range, counters and tables.

Receptacle Outlets. — One 20 amp outlet for refrigerator. Two 20 ampere circuits on counters. One outlet for each 1.5 m (4 ft) of counter work space, 110 cm (44 in.) above floor. Counter 0.3 m (1 ft) wide or wider — must have convenience outlet (220–3b and 210–25b).

Special Purpose Outlets. — Range, dishwasher, food freezer. The waste disposer may be on one of the two 20 ampere circuits required in the kitchen.

Bedroom

Lighting. — General illumination — ceiling, valance, cove or cornice — wall-switch controlled.

Receptacle Outlets. — Use the 2 meter rule. Outlet on each side and within 2 m (6 ft) of center line of each bed location.

Special Purpose Outlets. — Air conditioner

Bathroom

Lighting. — Provide illumination of both sides of face when at mirror (ceiling light also recommended), wall-switch controlled. Vapor-proof luminaire in enclosed shower stall, wall switch outside stall.

Receptacle Outlets. — One near mirror — 1 m (3 ft) to 1.5. m (5 ft) above floor.

Special Purpose Outlets. — Infrared heater — timer controlled. Ventilating fan, wall-switch controlled.

Stairway

Lighting. — Wall or ceiling outlets to light each flight, multiple-switch controlled.

Receptacle Outlets. — One at each landing for cleaning.

Hall

Lighting. — Sufficient outlets to light entire area, wall-switch controlled. Watch for irregularly shaped areas.

Receptacle Outlets. — One for each 5 linear meters (15 ft) of hallway — no point more than 3 m (10 ft) from an outlet.

Closets

Lighting. — If 1 m (3 ft) or more deep or floor area more than 1 m^2 (12 ft^2), outlet just above door frame controlled by wall switch, door switch or pull chain. (See Code Art. 410–8 for specific details.) Light 0.5 m (18 in.) from any combustible materials.

Laundry

Lighting. — At least one ceiling outlet, wall-switch controlled. Illuminate work areas, such as sorting, washing, drying, ironing areas.

Receptacle Outlets. — At least one on a 20 ampere circuit.

Special Purpose Outlets. — Automatic washer, clothes dryer (220-18).

Recreation Room

Lighting. — Good general illumination, wall-switch controlled — ceiling or wall fixture, cove, valance or cornice lighting.

Receptacle Outlets. — Use the 2 meter rule. Consider desk, radio, TV, projector, fan, etc.

Porch

Lighting. — At least one lighting outlet, wall-switch controlled.

Receptacle Outlets. — One for each 5 m (15 ft) of wall bordering porch or breezeway. Use weatherproof outlets if exposed to moisture (410-57). *For non-enclosed porch, outlet must be protected with ground fault interrupter* (210-8a).

Garage

Lighting. — At least one ceiling outlet, wall-switch controlled, for one or two car garage. Include exterior light and multiple-switch control if garage is separate from house.

Receptacle Outlets. — At least one for trouble light, battery charger, etc., others according to use of garage.

Special Purpose Outlets. — Automatic door opener. Outlet in center of ceiling. Also install low voltage push button at house entrance.

Exterior

Lighting. — One or more at each entrance, wall-switch controlled inside. Illumination of long approach walls, terraced or broken flights of steps, wall-switch controlled inside house.

Receptacle Outlets. — Weatherproof, 0.5 m (18 in.) above ground (See Art. 410-57). Must be protected by ground fault interrupter (210-8a).

Note on *small-appliance* branch circuits (from N. E. C. 210-16):

For small appliance load in kitchen, laundry, pantry, dining room, and breakfast room of residences, *two or more* 20 ampere (small-appliance) branch circuits shall be provided for all receptacle outlets in the rooms. There are to be no other outlets on these circuits. These branch circuits are in addition to general purpose and special purpose circuits.

REFERENCES

ANON. 1977. National Electric Code 1978. Natl. Fire Prot. Assoc., Boston.

SURBROOK, T. C. 1973. Residential Electric Outlet Planning Guide. Mich. St. Univ., East Lansing, Mich.

Index

Other AVI Books

AGRICULTURAL PROCESS ENGINEERING
3rd Edition   *Henderson and Perry*
AN INTRODUCTION TO AGRICULTURAL ENGINEERING
Roth, Crow and Mahoney
CEREAL TECHNOLOGY
Matz
COMMERCIAL FRUIT PROCESSING
Woodroof and Luh
COMMERCIAL VEGETABLE PROCESSING
Luh and Woodroof
CORN: CULTURE, PROCESSING, PRODUCTS
Inglett
DAIRY TECHNOLOGY AND ENGINEERING
Harper and Hall
DRYING CEREAL GRAINS
Brooker, Bakker-Arkema and Hall
DRYING OF MILK AND MILK PRODUCTS
2nd Edition   *Hall and Hedrick*
ELEMENTS OF FOOD ENGINEERING
Harper
ENCYCLOPEDIA OF FOOD ENGINEERING
Hall, Farrall and Rippen
FOOD ENGINEERING SYSTEMS
Vol. 1   *Farrall*
GRAIN STORAGE—PART OF A SYSTEM
Sinha and Muir
POULTRY PRODUCTS TECHNOLOGY
2nd Edition   *Mountney*
PRINCIPLES OF FARM MACHINERY
3rd Edition   *Kepner, Bainer and Barger*
PROCESSING EQUIPMENT FOR AGRICULTURAL PRODUCTS
2nd Edition   *Hall and Davis*
WHEAT: PRODUCTION AND UTILIZATION
Inglett